# Weather

# for

# the

# New

# Pilot

# WEATHER
## for
## the
## New
## Pilot

## TOM MORRISON

IOWA STATE UNIVERSITY PRESS / AMES

"It is not without trepidation that this subject is broached: when the safety of others is at stake, advice should not be too freely given."

—Jim McCollam, *The Yachtsman's Weather Manual*

© 1991 Iowa State University Press, Ames, Iowa 50014
All rights reserved

♾ Printed on acid-free paper in the United States of America

First edition, 1991
  *Second printing, 1996*

---

Library of Congress Cataloging-in-Publication Data

Morrison, Tom, (Thomas A.)
        Weather for the new pilot/Tom Morrison.—1st ed.
            p.         cm.
        Includes bibliographical references and index.
        ISBN 0-8138-1773-0.—ISBN 0-8138-1772-2 (pbk.)
        1. Meteorology in aeronautics. I. Title.
    TL556.M67         1991
        629.132′4—dc20                                                  91-19260

---

# CONTENTS

# PREFACE

**IT WAS THROUGH FLYING** and hangar-flying with my friends at the Victoria Flying Club in British Columbia that I came to realize how very little we, as new visual pilots, knew about the weather and how dangerous this ignorance could be. We received a comprehensive, if brief, course in meteorology at ground school and some degree of practical introduction to the weather in flight training. It was hardly possible, however, in the time available for us to become acquainted with the infinite variety of forms which the weather can assume and the full range of its possible effects on our flight. There was much which we had, perforce, to figure out for ourselves.

This book began in 1986 as an attempt to circumvent this problem. The result, which was submitted to Iowa State University Press in 1987, was a glib, short, and somewhat facile sketch of the weather, offering little in the way of originality and differing only slightly in form from many other aviation weather books. Mr. Bill Silag, Managing Editor, very properly returned the manuscript with instructions to do better. Perhaps it was the panorama of cloud-draped mountains seen from a Boeing 727 cruising above the Rockies that inspired the concept of the present book which emerged from the typewriter in 1988 and was accepted for publication in 1989.

My thanks are due to Mr. Bill Silag of Iowa State University Press for ensuring that the work reached its full potential and to the production staff for their customary high standards of work.

Note: Reference is made to Cessna aircraft as typical light aircraft. No comment on the Cessna Aircraft Co. or its products is intended or implied.

Tom Morrison
*Vancouver 1989*

# AUTHOR'S NOTE

Men outnumber women in the pilot population and therefore the masculine gender is used when referring to pilots in a generic sense. No sexist connotation is intended or implied.

# INTRODUCTION

**NEW PILOTS,** on winning their wings, are free to launch themselves, with few restrictions and little supervision, into the ocean of the sky. The sky is unforgiving to the unwary and takes no account of innocence or ignorance. For new private pilots, the aircraft is small and of feeble performance; their control of it is far from perfect; they cannot as yet fly by instruments; their understanding of the sky and of how it can affect their flight is meager.

The first three hundred hours of a pilot's career are the most dangerous. Student and low-time private pilots suffer a worse accident rate than any other part of the aviation fraternity. Roughly half of the accidents result from deficient operation of the aircraft; the other half results from a failure to interpret the condition of the sky. All pilots must enter this phase. Some of them pass through it successfully; many of them give up flying before doing so; a small proportion of the total numbers are killed in accidents. In all civilian flying there is a period, and quite a long one, when new pilots are restricted to visual flight, whether they intend to become instrument-capable or not.

There are few books, if any, which consolidate between two covers a compendium of information on the sky specifically for the new pilot and which offer guidance in assessing the feasibility of an intended flight with regard to the condition of the sky. There are many books and sections of books on aviation meteorology. Many of them merely repeat information which has been current for decades. The best of them are written by and for the instrument-trained pilots of high-performance aircraft. The sky conditions which are of concern to such pilots are quite different to those which affect the new visual pilot.

The purpose of this book is to bring together the various factors which combine to make the sky "flyable" or otherwise for new visual pilots at any given time, to present them for their consideration, and thus help them to reach a sound judgment whether to take off or postpone the flight, whether to fly on or retreat.

It is assumed that the reader has at least attended private pilot groundschool and is therefore familiar with the basic concepts and terminology.

The further purpose of this book is to reduce the accident rate among low-time pilots to the benefit of us all.

# Weather
# for
# the
# New
# Pilot

# 1

# The Nature of the Problem

## Mariners in a strange ocean

**KAILA WAS THE IHALMIUT GOD** of the sky. Chief among the gods worshipped by the Ihalmiut people of Arctic Canada, Kaila represented all the elemental forces of nature. Kaila was in no way a personal god. He was enigmatic, neither benign nor actively hostile. He cared nothing for the Ihalmiut or for anyone else. Kaila existed. Kaila was. Kaila was the sky.

Who has not stood on the seashore and watched the sea? Who has not seen it lapping gently on the sand or moving silently with the tide? Who has not seen it stirred to madness, bellowing its demented fury, hammering and gouging with terrible fists and claws at the very ground on which they stand? And who has not been mesmerized by its beauty, terrified by its power, and has not turned landward once more, humbled?

Yet what is it that stirs the surface of the sea but that other ocean of the air? Every human being yet born is a dweller on its shores and has heard the thunder of its surf, yet how little do we know of it! As long as humanity has possessed the ability for abstract thought — to muse, to speculate — some of us have looked upward and outward into it with an insatiable longing to cast forth from its shores. Within the past split second in our history, a very few of us have become mariners in its depths, have wrapped ourselves in it, and have flown.

Almost anything will float on water, from a wood sliver to a steel ship. With relatively meager knowledge and equipment we can sail on it or penetrate a little beneath its surface. The same is not true of flight. We must master the whims of a self-willed contraption called an aircraft. Once launched, we find that the familiar world, when seen from aloft, has become bland and strange so that we must steer our course above it by means of calculations and intricate devices. By no means least, we must judge the consistency and behavior of that ocean of air on which we would place our weight. Even if our aircraft is in faultless condition, even if our control of it is of birdlike perfection, even if our navigation from place to place is of the most extreme precision, neglect or incompetence in judging the "flyability" of the sky can have disastrous results.

The ability to judge the feasibility of flight in any given condition of the sky is not imparted to any great extent by conventional flight training. "Learning to fly" means learning to operate the aircraft and to fly it from place to place. Groundschool instruction touches on the effects of wind and air temperature on the aircraft's performance and teaches a smattering of meteorology. Private pilot navigation exercises tend to be flown on days selected for their fair weather, because the object of the exercise is to teach navigation with the least possible interference. Some instructors are as anxious as their students to avoid foul weather. The student absorbs, as if by osmosis, an idea of what conditions are "flyable" and what are not, often without really knowing why. The accident reports show that a simple lack of awareness of the effects of certain atmospheric conditions is the cause of some very serious accidents.

New private pilots find themselves in an environment that is suddenly strange and uncertain. They are free to fly wherever and whenever they wish, with or without passengers, at any time of day or night, during all seasons of the year, and over any type of terrain. The legal limits of the Visual Flight Rules include some really foul flying weather and make no distinction between flight over the mountains on a winter night and flight over the prairies on a summer day. The Visual Flight Rules offer no safe guidance.

If pilots rent aircraft from a Fixed-Base Operator (FBO), they are subject to some restraint in that the FBO may refuse to dispatch a flight into certain atmospheric conditions. If they buy their own aircraft, they are not subject even to those restraints. No one is able or willing to advise pilots directly whether or not to launch a flight, and new pilots are not always sure how to assess the situation for themselves. The weather office may have given them a string of information that they are barely able to interpret.

New pilots are told that the decision is theirs alone. This is an immutable truth, but the necessary information on which to base that weighty decision is often lacking. We might as well ask a group of civil engineers to design a bridge without telling them what loads it must bear or what materials are available for its construction.

New pilots are further confused by the antics of their fellow aviators. They see airliners thundering into the gloom, vanishing into clouds and rain before the wheels are in their wells. They see other lightplane pilots taking off into conditions they themselves cannot accept, and wonder why they alone are left on the ground. Their decisions may swing wildly between excessive and insufficient caution, especially when pressures are upon them to reach a destination by a certain time or to satisfy the needs of passengers.

Unfortunately some of these situations result in accidents. A private pilot with 107 hours damaged his aircraft landing in a crosswind in wind speeds of 25 knots gusting 35. A private pilot with 65 hours damaged his aircraft when it failed to take off uphill, fully loaded, on a hot, windless day at a grass strip 2,700 feet above sea level. A private pilot with 112 hours damaged the aircraft in a forced landing caused by carburetor ice. A private pilot with 230 hours was flipped over while taxiing in a 35-knot wind. A private pilot with 137 hours managed to survive a forced landing in whiteout conditions in the mountains. A private pilot with 148 hours was injured in a crash following disorientation while flying circuits at night. A private pilot with 105 hours damaged his aircraft and injured himself in a precautionary landing. The accident report stated: "The pilot had obtained a weather briefing prior to flight but apparently did not appreciate the fact that flight over ascending terrain in forecast low ceiling, fog, and snow showers would be inadvisable. On encountering deteriorating weather, he pressed on until his options were limited." The weather conditions at that time included a 20-knot wind, 200-foot overcast ceiling, ½-mile visibility, and a temperature just below freezing.

These accidents, pitiful and unnecessary as they were, are nothing compared to the tragedy of fatal accidents, especially those involving passengers who had unwittingly placed their lives in the pilot's hands. Two private pilots, the pilot-in-command having 133 hours, flew Visual Flight Rules (VFR) in the mountains without mountain experience, when tired, at night, in weather conditions below VFR minima: two fatalities. A private pilot with 68 hours crashed in whiteout after taking off into 2-mile visibility in snow: one fatality. A private pilot with 188 hours took off in a high-performance, single-engine aircraft with no weather briefing; he lost control of the aircraft at dusk or at night in heavy snow and low cloud: six fatalities. A private pilot with 142 hours lost control of his aircraft in snow and bad visibility over a frozen river: two fatalities. A private pilot with 228 hours crashed in fog: four fatalities. It was reported: "The FSS briefed that IFR, improving to marginal VFR conditions with low ceilings and rain showers, could be expected enroute and at destination. Despite this, the pilot set out with three passengers on a VFR flight." A private pilot with 80 hours and his two passengers were killed when his aircraft crashed into the sea in marginal VFR conditions. A student pilot was flying circuits under a 650-foot overcast ceiling with 4 miles visibility and a windspeed of 18 knots. He entered clouds inadvertently, went into a spin, and was killed in the ensuing crash.

And so the gruesome tale goes on. In all cases the retrospective view is that the pilots should not have been flying "in those conditions." The pilots were, we must assume, reasonably competent individuals who had, after all, demonstrated that they were fit to be in charge of an aircraft. In essence, they lost control over the progress of their flight because they took off into or entered atmospheric conditions beyond the aircraft's performance limits or beyond their skills to handle. Either they were unaware that such conditions were hazardous to them, or that such conditions existed at the time, or both. In the case of student pilots, this unawareness evidently extended to those who dispatched the flights.

Over the past 40 years the airlines, corporate aviation, and the military have made great strides in reducing their accident rates. This improvement has not been matched in private aviation. In a society that is safety-conscious to the verge of absurdity, this is a serious matter. Few countries specifically outlaw private flying, yet there are few countries where it is feasible. The answer to this apparent paradox is that in most countries of the world where the individual's disposable income suffices to cover the basic costs of private flying, the regulations are so burdensome and compliance with them is so costly that private flying has been effectively legislated out of existence. In Canada, where many of the air regulations begin: "No person shall operate an aircraft unless . . . ," the private pilot population has been shrinking during the 1980s at a rate that could lead to extinction in 25 years. In an intensely urban society such as North America, an element of the population believes that the only legitimate reason for an individual to fly is for transportation, and that those needs are adequately met by the licensed air carriers. These people would argue that personal flying for recreation or business is unnecessary and *for that reason* should not be allowed. Regardless of whether opinions such as these carry any real weight, the immediate and pressing concerns of insurance costs and compliance with increasingly stringent regulations present a problem.

Private flying is a precious privilege, and we should be under no illusions as to the ease with which it can be eroded and finally taken away. The private pilot population must show itself worthy of this privilege beyond doubt. Accidents related to atmospheric conditions form a substantial proportion of total general aviation accidents, and anything that attacks this cause will have a major effect on the accident rate as a whole.

This book seeks to bring together as many as possible of the atmospheric variables that combine to affect the feasibility of visual lightplane flight. Thus, day, night, terrain, and population density are considered just as important as the height and type of

clouds. The causes and effects of disorientation and the importance of ceiling and visibility are looked at here more closely than
in some other books of the same general nature.

The atmospheric variables are divided into two categories:
those affecting the aircraft's performance and controllability, and
those affecting the pilot's ability to see. Each section is further
subdivided between those phenomena that change frequently and
to some extent unpredictably (the weather) and those that change
either with absolute regularity or not at all (the diurnal and
seasonal cycles and the terrain).

The only certain thing about the sky is that it will change.
The questions are: How? How much? How soon? The ever-
changing nature of the sky is what gives it its subtlety, its beauty,
and its endless fascination. To predict and interpret its moods for
the purposes of flight is one of the highest skills to which a pilot
can aspire.

# 2

# Temperature
## The heat of the day and its absence

**THE DRIVING FORCE** of all our climates and weather is the sun's heat. The flow of heat from the earth's interior is negligible in this respect. The way in which the sun's heat reaches the earth, and what happens when it does, is surprisingly complicated.

## Solar Radiation

Solar radiation is distributed unevenly over the earth's surface. We all know that the earth spins on its axis, exposing each place on its surface to a 24-hour cycle of day and night. We also know that the axis is tilted at an angle to the direct line from the sun, and that this tilt angle remains constant in space regardless of the earth's annual circuit around the sun. In the northern summer the North Pole leans toward the sun and the South Pole away from it, whereas in the northern winter the opposite is the case. This causes the annual cyclic variation in solar radiation reaching any place on the earth's surface, which defines the seasons.

One of the most important, and least appreciated, facts about the atmosphere is that it is heated and cooled almost entirely by contact with the earth's surface and is heated only slightly by the sun itself. This heating and cooling has immediate effects on the lowest 10,000 feet of the atmosphere, and the ways in which it occurs are important to us.

Large-scale global wind circulations occur at all levels in the atmosphere. These are driven by seasonal heating and cooling of the earth's surface, but they do not directly concern us here.

Of the total incoming solar radiation, it is estimated that 3% is absorbed by the stratosphere, 15% is absorbed by gases and dust in the troposphere (the lowest part of the atmosphere), 35% is reflected back into the stratosphere, and 47% reaches the earth's surface. Of this 47%, one-third is radiated back into space, one-third is absorbed in evaporating water, and one-third heats the lower atmosphere.

Different surfaces reflect different percentages of the solar energy that strikes them. This reflectivity, called the albedo, is the percentage of energy that is reflected. Typical values are:

| | |
|---|---|
| Fresh snow, low sun | 90–95% |
| Fresh snow, high sun | 80–85% |
| Thick cloud | 70–80% |
| Cloud layers more than 3,000 ft thick | 45–85% |
| Old snow | 50–60% |
| Water, low sun | 50–80% |
| Thin cloud | 25–50% |
| Cloud layers less than 500 ft thick | 5–65% |
| Sand | 20–30% |
| Grass | 20–25% |
| Dry earth | 15–25% |
| Wet earth | 10% |
| Forest | 5–10% |
| Water, high sun | 3–5% |
| Planetary average | 30% |

We can see at once that the solar energy finally available to heat the earth's surface, and hence the bottom of the atmosphere, is highly variable.

## Heating Rates and Cycles

Land and water heat up and cool down at different rates. The specific heat of water is five times that of rock or dry soil, and three times that of wet soil. That is to say, five times as much heat energy is required to raise the temperature of 1 pound of water by 1° of temperature as is required to heat 1 pound of rock by the same amount. Thus if the sun shines with the same intensity on a rock surface as on a water surface, the rock will become much hotter than the water. Differences in specific heat, however, are not the only reason why this is so.

On land the sun's energy is almost completely intercepted by the first few inches of soil or rock so that at a depth of 3 feet in soil, the temperature remains constant throughout the 24 hours. At a depth of only a few feet in solid rock, the temperature is absolutely constant year-round. The surface temperature, by contrast, can fluctuate by a range of 20°C/36°F between day and night; exceptionally, this range may reach or exceed 30°C/54°F.

In the case of water, solar energy penetrates beneath the surface to the extent that 10%–40% may penetrate to a depth of 3 feet. Water also can be mixed by waves and currents, with the result that incoming energy is dispersed to greater depths. Consequently, the daily temperature range at the surface varies by less than 1°C/2°F. This figure may reach 10°C/18°F in shallow water because of heating of the sand, mud, or rock bottom by solar radiation passing through the water. Once the sun's heat is gone, land surfaces radiate heat into space quicker than do bodies of water, whether at night or in winter. Heat flow from the earth's interior has no measurable effect on weather or climate.

Therefore, by day or in summer, the land heats up while the water remains cool. At night or in winter, the water remains relatively warm while the land becomes cold. Even a large lake can affect local climate and weather. Most of us know the extremes of temperature experienced in continental areas far from seas or large lakes, as compared to the more moderate temperatures experienced in coastal areas. The Great Plains, for example, may have a greater range of temperature in a day than some coastal areas do in a year.

Because the lowest part of the atmosphere is heated and cooled by conduction of heat to or from the earth's surface beneath it, these contrasts are of great importance to many weather forms and we will refer to them repeatedly. Figure 2.1 illustrates diurnal variations in air temperature near the ground.

Of the solar radiation the earth receives, a good proportion is reflected and radiated back into the atmosphere. But the same mechanisms that absorb and reflect the sun's energy inbound through the atmosphere also intercept energy outbound from the earth and reflect or radiate some of it back down again. Thus, although a cloud deck prevents a substantial proportion of solar energy from reaching the earth, it also traps energy radiated by the earth, with the result that a cloudy night is often warmer than a clear one. An additional complicating factor is that solar energy consists of a spectrum of wavelengths, including visible, ultraviolet, and infrared light. Energy of different wavelengths behaves in different ways and is absorbed, reflected, and reradiated by different materials at different rates.

Other factors affecting the temperature range and cycle at the earth's surface at any one place are wetness of the surface and its vegetation, type and thickness of the vegetative cover, lightness or darkness of the surface, type of surface (forest, grassland, soil, sand, rock), passage of precipitation, steepness and aspect of slopes, and strength of the wind.

We can see that the energy exchange between the sun, the earth, and the atmosphere is far from simple. What we have just looked at is by no means the full story, but it will have to do for now.

There is a balance between incoming solar energy and outgoing radiation from the earth both between day and night and between winter and summer. In simple terms (for a subject that is far from simple!) by day the incoming radiation from the sun exceeds the outgoing radiation from the earth so that the earth's surface becomes warmer; at night, in the absence of solar radiation, the earth continues to radiate heat into space and so becomes colder. The highest surface temperatures are reached in the middle of the afternoon, the lowest just before sunrise.

In winter, days are shorter and the sun is lower in the sky than in summer. When the sun is low in the sky, its rays must pass

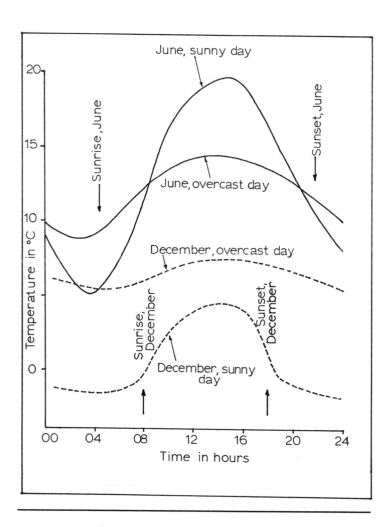

Fig. 2.1. Examples of diurnal variations in air temperature near the ground (air temperatures at 1.2 meters above ground level). Adapted from R. Geiger, *The Climate Near the Ground* (Cambridge, MA: Harvard University Press; originally published in 1927).

more obliquely through the atmosphere and more of its energy is intercepted before reaching the earth. Therefore, during the winter there is a net loss of heat from the earth's surface and the lowest temperatures occur toward the end of winter in January and February. During the summer, by contrast, there is a net gain of heat, and the highest temperatures occur in late summer in July and August.

Large masses of air are conditioned by seasonal heating and cooling of the earth's surface. This has marked effects on weather, but we will come to that later on.

## Standard Atmosphere

As we venture into the sky, we need an array of reference data. One of these is the *standard atmosphere.* Actually, there are three standard atmospheres, all of which are the same at aviation altitudes and all of which represent average conditions at about 45° North latitude. These are the American Standard Atmosphere 1976, the International Civil Aviation Organization (ICAO) Standard Atmosphere, and the International Standards Organization (ISO) Standard Atmosphere. According to these standards, the sea-level air temperature is 15°C/59°F, decreasing at a *lapse rate* of 1.8°C/3.2°F per thousand feet of height. The actual lapse rate often varies widely from the standard lapse rate, just as the sea-level temperature often varies widely from 15°C. We will come to lapse rates again later on. An aircraft's performance figures are predicated on standard conditions and when air temperatures vary widely from standard there are effects of which we must be aware:

1. the effects of heat and cold on the aircraft's engine, fuel, and lubrication systems;

2. the effects of heat in degrading the aircraft's takeoff, climb, service ceiling, and landing performance; and

3. the effects of a combination of cold and moisture to produce ice in and on the aircraft, and the effects of snow and ice on airfields.

## Effects of Temperature

The effects of heat and cold on the aircraft's power plant and related systems vary in detail from one type of aircraft to another. These details are beyond the scope of this book; they are described in operators' manuals. When air temperatures rise above 25°C/80°F or fall below freezing, the manual should be consulted. Extremes of heat and cold are characteristic of the sparsely settled areas of the world. The reader is referred to the chapter on "Northern Wilderness Flying" in *Van Sickle's Modern Airmanship,* and the chapter on "Desert Flying" in Barry Schiff's book *The Proficient Pilot II.*

The main problem in hot weather is overheating of the engine, which can occur if the aircraft is not moving fast enough to provide adequate cooling airflow over the cylinders. This can happen if the aircraft spends too long taxiing in calm, hot air or facing out of wind, or if the runup is prolonged unnecessarily and is carried out with the aircraft facing out of whatever wind there may be. A long full-throttle climb at best-rate-of-climb speed also can bring the oil temperature to the red line. Keeping a higher airspeed in the climb will prevent this from happening. Some fuel systems have been prone to vapor locking in very hot weather. If the engine oil is subjected to temperatures above its design limit, it may break down and fail to lubricate the engine properly.

Oil can also cause problems in low temperatures. If oil of the wrong viscosity is used, or if it is allowed to become excessively cold, it may be too thick to circulate through the engine. Batteries are weak at low temperatures, and at the same time the engine is harder to turn over because of the viscosity of the cold oil and the tightened tolerances caused by the metal contracting in the cold. There is a risk of thermal shock to the engine unless it is warmed up gradually before takeoff.

If the air is cold, a long descent with the engine idling is damaging to the engine. A jet of cold air from the cabin heater should warn the pilot of this condition. A number of light aircraft accidents have resulted from a long power-off descent, either in a normal descent to land or during practice forced approaches. In such a situation the engine cools excessively, with or without the formation of carburetor ice, which the engine then may be too cold to remove by means of carburetor heat. As a result, the engine does not respond when the pilot reapplies throttle, and either the aircraft fails to reach the runway or a simulated forced approach becomes a real one. The colder the air, the greater is this hazard.

The ideal solution is a continuous power-on descent from cruising altitude to final approach. This has the additional benefits of giving passengers the smoothest possible ride (which impresses them accordingly) and of moving the aircraft quickly through the airport traffic system. Conversely, the worst method is a descent in a series of steps leaving the aircraft at a low altitude far from the runway. At night this can be dangerous as well.

As a rule, "Cold Weather Operations" are in force when the temperature is below freezing, and the aircraft manual should be consulted accordingly. Extreme cold, for aviation purposes, means temperatures below $-18°C/0°F$; in such conditions special procedures are needed. At temperatures below $-15°C/+5°F$, moisture in the nostrils can be felt to freeze with each indrawn breath, giving warning of these conditions.

Fuel itself does not freeze at temperatures normally encountered on earth, but gasoline contains small amounts of water in solution as tiny droplets. As fuel is heated, its capacity to hold dissolved water increases and it will absorb water vapor from the atmosphere. As the fuel cools, its capacity to hold water decreases and the water precipitates out. This is the water that appears in the sampling cup as globules when we drain the fuel tank sumps on a cold, damp morning. Water droplets suspended in the fuel can freeze. In liquid form they would pass harmlessly through the engine, but as ice crystals they can plug fuel filters, bends in fuel lines, valves, or the carburetor jet. Anti-icing compounds are premixed with some fuels. If water is allowed to accumulate in the tank sumps and freezes, it cannot be drawn off through the drains. If the aircraft then flies into warmer air, this ice will melt and the resulting water may contaminate the fuel.

New pilots will become familiar with operating procedures in their home territories. But when New Englanders fly in Arizona in the summer, or Floridans in Minnesota in the winter, they must be aware that the different temperatures will affect their aircraft and they must ensure that their aircraft is in fit operating condition. One thing not readily apparent is that in some aircraft the engine-cooling baffles and oil cooler may be arranged so as to work properly within a certain temperature range. These arrangements may have been made locally and will not be apparent from the preflight inspection. That is another reason why it is important to check out an aircraft thoroughly before flying in temperatures markedly different from those at its home base.

## Density Altitude

The air temperature produces an effect known as *density altitude*. Air consists of molecules of its constituent gases, which are in a constant random motion called *Brownian motion*. The hotter the air, the more the molecules move about. In fact, the definition of the temperature at which there is no heat at all (absolute zero, 0° Kelvin, −273°C) is the temperature at which Brownian motion ceases. If air is heated inside a closed container, the molecules beat on the inside of the container more and more violently and the pressure rises. If unconfined air is heated, it expands as the molecules collide more and more violently with one another, and its density—the number of molecules per cubic inch—decreases.

Molecules of the air's constituent gases are what the propeller bites on and what supports the wings in flight. If we imagine ourselves walking across a pond on stepping stones, we can see that, if the stones are close together, we can move slowly. The farther apart they are, the faster we have to move to stay on the stones. The analogy is not exact, but it suffices for the present. The higher the temperature, the faster the wing must move—that is, the higher its true airspeed must be to find enough molecules to bear the aircraft's weight.

The propeller becomes less effective because it, too, is an airfoil. It drags the aircraft forward by accelerating air backward. If the molecules are relatively few and far between, fewer of them are contained in the arc swept by the propeller, which produces correspondingly less thrust.

If the air is thinner, the propeller must meet less resistance, so why does it not just turn faster and so find the same number of molecules to accelerate backward as before? Because the power of the engine also is reduced by the lower air density. The engine derives its power from the gasoline. The gasoline contains a certain number of units of energy per pound, called its *calorific value*. The more gasoline burned per minute, the higher the power output. But that presupposes the most efficient possible combustion of the gasoline, molecule by molecule, for which the right amount of air has to be present. If each cubic foot of air drawn into the cylinders contains relatively few molecules, either the gasoline is incompletely burned and goes out through the exhaust unburned or as soot, or the pilot leans the mixture, which reduces the gasoline flow rate and so readjusts the gasoline/air mixture to the right proportions. In either event, the number of pounds of gasoline burned each minute is reduced and the engine loses power.

Density of the air also decreases with increasing altitude above sea level. Thus, the air at Denver, 5,300 feet above sea level, is less dense than the air at San Francisco when the temperatures are the same. Because of the effect of temperature on density, however, the air at San Francisco at 30°C/86°F actually would be of the same density as the air at Denver at −20°C/ −4°F. This combined effect of temperature and altitude can be expressed in terms of an altitude equivalent called density altitude. In both of these instances, the density altitude is 2,000 feet, and an aircraft would perform at either place as if it were flying 2,000 feet above sea level in the ICAO Standard Atmosphere. Knowing this, we could consult the aircraft's manual and predict takeoff roll, initial rate of climb, and landing roll.

Variations in barometric pressure also affect air density: The higher the pressure, the more the molecules are squashed together and the more of them there are in each cubic foot of air. To be perfectly accurate, we should have said: "An aircraft would perform as if it were flying at a *pressure* altitude of 2,000 feet." *Pressure altitude* and *true altitude* are the same when the sea-level barometric pressure equals the ICAO standard sea-level pressure of 29.92 inches of mercury, or 1,013.3 millibars. The effect of barometric pressure on air density is, however, slight. At 5,000 feet above sea level, the air pressure is about 850 millibars. In comparison, sea-level pressures as low as 920 millibars have been measured in the eyes of hurricanes. The whole range of normal sea-level barometric pressures is equivalent to the difference in air pressure over an altitude range of about 1,000 feet. Five hundred feet is about as close as we can read on the density altitude scale of an E6B computer. Therefore, for most purposes, pressure and true altitudes can be regarded as the same.

Temperature and altitude both affect air density, and air density is critical to flight. We know that, as we climb, the air temperature decreases. We also know that the ground heats the air, whatever its elevation above sea level. The temperature at ground level can be 15°C/60°F in Seattle and 27°C/80°F in Denver at the same time on the same day, even though one is at sea level and the other is 5,300 feet above. The density altitude at the two places would be very different—0 feet at Seattle and 7,000 feet at Denver. An aircraft's takeoff and landing performance would correspond to these density altitudes regardless of the actual runway elevations.

Those of us who fly from long runways at sea level in the temperate or cold parts of North America may never encounter density altitude as a significant problem. Should such pilots fly in the mountainous West in the summer, however, they could be unpleasantly surprised. So, too, could pilots living in the mountains who are too careless or incompetent to take density altitude into account.

The problem is that the higher the density altitude, the higher is the true airspeed to which the aircraft must accelerate to lift its load. At the same time, the means of obtaining the necessary acceleration—the propeller—produces less thrust. Therefore, a high density altitude results in a long takeoff roll and a sluggish climb.

On landing when the density altitude is high, the wing loses its ability to support the aircraft while its true airspeed is still relatively high. Therefore, the touchdown speed is high and the landing roll is long.

In general, if a light aircraft is flown correctly, the landing distance between crossing a 50-foot obstacle and coming to a stop is similar to, but less than, the takeoff distance needed to clear a 50-foot obstacle. Partly for this reason, and partly because of the impact speeds involved, accidents caused by failed takeoffs at high density altitudes are more common and more serious than those involving landing overruns in such conditions. Nevertheless, pilots who habitually fly at excessive speed on final approach could be in for trouble if they were to land on a short, high-altitude runway on a hot day.

The region of real concern is the mountainous West of Canada and the U.S., where high surface elevations and heat in summer combine. Airports such as Denver Stapleton have ample runways, but then we encounter places such as Aspen, Colorado, at 7,800 feet with a 7,000-foot runway sloping at 2%; Leadville, Colorado, at 9,900 feet with a sloping 5,300-foot runway; and many others above 4,000 feet. Summer temperatures can reach 25°C/80°F, and density altitudes can restrict takeoffs to the early morning and late evening hours.

There is all the difference in the world, too, between the takeoff performance of a lightly loaded Cessna 182 or equivalent aircraft and that of a fully loaded Cessna 172 or equivalent. For example, a Cessna 182 with fuel tanks half full and only the pilot on board has a power loading of only 9.6 pounds per horsepower, whereas a Cessna 172 at gross weight has a power loading of 14.4 pounds per horsepower, or 50% greater. Each available horsepower has 50% more load to move. The strutted high-wing Cessnas all look much alike, as do the single-engine, four-seat Pipers. A new pilot with a fully loaded Cessna 172 might see a lightly loaded Cessna 182 take off successfully at a high density altitude, but he could be heading for a serious accident if he were to try to do the same.

In the dry air of the West, heat and cold are less noticeable than in the humid East. The temperature might be a fiery 40°C/100°F with attendant density altitude effects while the air feels more like 25°C/80°F.

Even the aircraft operating manual requires careful interpretation. Let us suppose that we are considering a flight in a fully loaded Cessna 172N from a place in Colorado at 7,700 feet elevation through a pass at 10,400 feet, and that the temperature is 25°C/80°F. The manual tells us that in those conditions we should be able to clear a 50-foot obstacle in 3,250 feet at 8,000 feet and 25°C. The runway is 7,000 feet long, so that appears to be fine. The aircraft's service ceiling is given as 14,200 feet, so that appears to be fine as well. There are, however, certain snags which could result in serious consequences.

The figures in the manual were obtained with a new aircraft flown by a test pilot. An old aircraft flown by a new pilot will not match the performance of a new aircraft flown by an old pilot. The extent of the shortfall is indeterminate.

When the aircraft's takeoff performance is marginal, it may become airborne in ground effect but be unable to climb further. Ground effect facilitates flight within one wingspan's height above the ground (by rule of thumb) because of the interference between the ground and the air circulation around the wing. (When certain kinds of sea birds fly in V-formation a foot or two above the water, they are not only flying in water effect but also are riding each other's wingtip vortices, which minimizes the effort needed to fly.) The pilot is likely to discover that he cannot climb out of ground effect when a safe landing in the remaining length of runway is no longer possible. In practice we cannot predict the density altitude at which this condition will set in, which suggests leaving an ample margin.

By using our E6B, we can see that at an elevation of 8,000 feet and a temperature of 25°C/80°F, the density altitude is 11,000 feet. From the manual we can see that at 11,000 feet the aircraft will supposedly climb at 250 feet per minute. Therefore, our initial climb rate should be 250 feet per minute. At best-rate-of-climb speed, 250 feet per minute is only 250 feet per mile, which is not very much. Downdrafts or subsiding air could reduce this still further or wipe it out. Moreover, that 250 feet per minute is obtainable in straight flight at exactly $V_y$. Any turns or deviation from $V_y$ will reduce it. We face a serious risk in that we may be unable to outclimb or avoid obstacles or that downdrafts may push us back to the ground. It is not a promising situation.

Even less promising is our intended flight through the pass. At 25°C/80°F the density altitude in the pass is about 14,000 feet, and we cannot attempt it. An aircraft close to its service ceiling is in a wallowing, mushing condition of flight and in no condition to handle the turbulence common in mountain passes.

Sometimes a light aircraft will not even reach the service ceiling predicted by the manual. This could be the result of improper operation, an aging engine, or large-scale subsidence of air such as takes place in anticyclonic conditions. The reduction in ceiling is extraordinary in some cases. A Cessna 172, climbing southbound out of Medford, Oregon, at gross weight one August day, was unable to climb above 8,000 feet. A thermal rising from a barren peak on the California state line pushed it up to 9,500 feet, where it stayed. As the fuel load lightened and the sun sank in the west, it drifted up to 11,000 feet over the Sierra Nevada and would have flown higher had the pilot not descended to land at Lake Tahoe. The aircraft's actual ceiling varied according to fuel load and atmospheric conditions by 25%.

It is probably reasonable to say that takeoff in a fully loaded average light aircraft should not be attempted if the density altitude is higher than 6,000 feet or that altitude at which, according to the manual, the rate of climb falls below 400 feet per minute. It could be argued that a brisk headwind would help the aircraft to take off, but that same wind would produce turbulence, and possibly downdrafts, in the country surrounding the airfield. Four hundred feet per minute is a decidedly anemic rate of climb with which to tackle turbulence and downdrafts, particularly if turns are necessary immediately after takeoff. If possible, turns should be toward sunlit ground, which may produce thermal lift, and not toward shaded ground, where air is more likely to be sinking.

The main point to bear in mind is that in many situations of high density altitude, the aircraft's performance is likely to be seriously degraded just when it is most needed.

## Effects of Combined Cold and Humidity

Having looked at the effects of heat, we now should consider the effects of cold and humidity combined. The chief hazard is the formation of ice in and on the aircraft. That well-known product of the sky, snow on the ground, is something we should not forget.

Carburetor ice, when to expect it and how to avoid it, is dealt with so copiously in basic training that it is unnecessary to go into any detail here. Carburetor ice is almost certain to occur in cool, damp weather when the engine is running at low speed. It will be noticed soon after starting the engine, or while taxiing, when the engine begins to choke and falter. It would also be noticed, with embarrassing results, during a descent in cold weather if carburetor heat were not applied or if, by some misfortune, the carburetor heat control cable were to break. Because the temperature may drop as much as 22°C/40°F in the carburetor throat, carburetor ice can form in ambient temperatures as high as 21°C/70°F if the air contains enough moisture. (A *range* of 22°C is equivalent to a *range* of 40°F. A *temperature* of 21°C is equivalent to a *temperature* of 70°F. Hence the apparent discrepancy between "22°C/40°F" and "21°C/70°F.")

Ice forms more readily when the throttle is partly closed because of a greater pressure drop across the carburetor throat, and hence a greater temperature drop, than when the throttle is fully open. Most people who have used compressed-air machinery in cool or cold conditions will have noticed ice forming around the exhaust. The sudden pressure drop as the compressed air escapes to atmosphere causes an equally sudden temperature drop and condensation and freezing of the moisture in the air. The same thing happens in the carburetor throat, assisted by the fact that the evaporating gasoline is also drawing its latent heat of evaporation from the air.

Some people, myself among them, have wondered why carburetor ice is not a problem in automotive engines. The answer is that carburetor heat is supplied continuously through a length of trunking leading from a shroud over the exhaust manifold to the carburetor air filter. Continuous carburetor heat is undesirable in airplane engines because of the slight loss of power caused by feeding hot air into the carburetor. (We see this in an airplane engine through the drop in rpm that occurs when we apply carburetor heat.) If the carburetor heat trunking in an automotive engine becomes disconnected, carburetor ice will form in the same conditions and with the same effects as in an airplane engine.

Ice can also form on the outside of an aircraft in flight. Cloud droplets remain liquid at temperatures well below freezing unless they are disturbed, as by impact with a solid object. This is all too familiar to inhabitants of the northeastern U.S. and eastern Canada when freezing rain falls and coats everything with ice.

Ice will form on the leading edges of an aircraft in flight if there is visible moisture — cloud, drizzle, rain — and if the outside temperature is between $0°C/32°F$ and $-15°C/+5°F$. Ice accretion has been encountered in temperatures as high as $+3°C/+8°F$ and as low as $-40°C/-40°F$. Ice forms on wings, propeller blades, stabilizers, and antennas. It can block pitot and static vents, air intakes, and especially the vacuum venturis seen on some older aircraft.

Ice accretion destroys the aerodynamic shape of whatever it forms on, so reducing the thrust of propellers and the lift of wings. It can cause dangerous vibration in propellers by unbalancing them, and can break antennas off. Ice accretion increases parasite drag. As a result, the airframe offers more resistance to airflow at the same time that the propeller is producing less thrust and the wings less lift. The combined effect is a rapid loss of airspeed. The wing must be flown at a higher angle of attack to sustain level flight and thus offers more frontal surface area on which ice can form. At the same time, the stalling speed increases and the aircraft's stalling characteristics can become vicious and unpredictable. This can make the aircraft more liable to stall and spin, and difficult or impossible to recover. Blockage of pitot and static vents causes false readings of the altimeter, airspeed indicator, and vertical speed indicator. Fortunately, ice accretion is not a common problem for visual pilots because they do not fly in clouds. On the other hand, freezing rain and freezing drizzle are extremely dangerous and must be avoided.

To try to take off with ice, frost, or snow on the wings is dangerous. These alter the airflow over the top surface of the wing, degrading its performance and altering its stalling characteristics. Laminar flow wings are especially vulnerable. Snow will not "just blow off." On closer inspection it may be found that the wing was relatively warm when the snow began to fall, with the result that the snow melted. As snow continued to fall, or night came on, or a cold front passed, the air cooled and the partly melted snow refroze to form a hard, tenacious honeycomb of ice. It was only the subsequent snow that did not melt but instead formed a powdery deposit. Snow or frost, no matter how thin, must be cleaned off the wing before flight is attempted. In the absence of a heated hangar or antifreeze, fuel from the sump drains will remove ice.

Plenty of aircraft, large and small, have come to grief in strong winds, especially crosswinds, blowing over icy runways and taxiways. In some ways a light aircraft is in a better position than a large one, because its momentum on touchdown is less and it is smaller in relation to the dimensions of most runways. For taildraggers an icy runway increases the risk of groundlooping.

Snow on the airfield is a fruitful source of accidents, usually minor ones. Landing on a snow surface is risky if the pilot does now know the depth or consistency of the snow. Snow can conceal holes, ditches, soft spots, and small obstructions. In dim light the absence of shadows and contrast can conceal snowbanks and irregularities in the snow surface. The amount of power needed to taxi through even shallow snow is surprising, and the aircraft can easily become bogged. Wheel pants are removed for winter operations because of the risk that slush will freeze inside them and lock the wheels. Half an inch of slush or wet snow on the runway can double the takeoff roll; an inch can prevent takeoff altogether. Packed snow offers a hard but rough surface. The new pilot who has not operated from a snow-covered airfield will find it worthwhile to obtain actual dual instruction, and not merely a briefing, from one who has.

Air temperature has many and varied effects on an aircraft's performance. The more widely the temperature diverges from ICAO standard, the more marked the effects will be.

# 3

# Wind
## Blowing
## where
## it
## wants

**WIND IS AIR MOVING.** It profoundly affects our lightplane operations. It may of itself determine the feasibility or otherwise of our intended flight. It can help us to forecast the weather.

We can usefully divide winds into two kinds: regional and local. *Regional winds* are the movement of air in response to patterns of barometric pressure hundreds or thousands of miles across. *Local winds* spring up and die down in response to local variations in temperature and because of convection.

## Regional Winds

Regional winds are the whole sky on the move. They are likely to persist for periods of at least 12 hours. They may blow for days almost independently of day and night. They are unlikely to change rapidly in strength or direction except with the passage of a front.

Regional winds are forecast in the area, terminal, and winds aloft forecasts issued by aviation weather offices (see Chapter 12). There is a marked correlation between light regional winds and good VFR flying weather. We think of such weather as being "settled." Winds at 6,000 and 9,000 feet of 15 knots or less are typical of these conditions. They often are associated with anticyclones or ridges of high pressure with barometric pressures higher than 1,015 millibars or 30.00 inches of mercury.

The stronger the regional wind, the more and the sooner the weather is likely to change. A strong regional wind is bringing large volumes of air from far away, perhaps over a wide variety of land and water surfaces. Therefore, the contrast between the temperature of the air and that of the land or water beneath it is likely to be enhanced, and this has marked effects on atmospheric stability. Any orographic lifting is that much more violent in strong winds. Fronts approach more quickly and are more active when they arrive than if the regional winds are lighter.

The thermal phenomena that cause local winds are usually at their weakest in the morning and evening. Therefore, the wind blowing at those times is most likely to be the regional wind. If this wind is stronger than a fresh breeze, changeable weather may be expected. But we cannot say, conversely, that if the regional wind is light, the weather will not change in the next 12 hours. Local effects such as fog, low stratus, or thunderstorms are possible. Regional winds are likely to persist independently of day or night as long as they are strong enough to suppress local thermal effects. Winds at 6,000 feet and above are most likely to be the regional winds.

Winds of 15 knots or less at cruising altitude can safely be ignored in flight planning unless the wind is a headwind and fuel supplies are critical. A good rule of thumb, however, is to allow for a 10-knot headwind on all long westbound flights.

Jetstreams are commonly the affair of the high-altitude jet pilot because they occur near the tropopause (see Chapter 9) at heights on the order of 30,000 feet. Low-level jetstreams, however, have been identified at altitudes as low as 1,000 feet above ground level at night over the Great Plains. Wind speeds may reach 60–70 knots. A lightplane pilot encountering such a jetstream might be warned by turbulence around its margins. If flying upwind in the jetstream, his groundspeed would be so reduced that he could run out of fuel enroute if he were not keeping a close check on his progress.

## Local Winds

Local winds tend to persist for periods of less than 12 hours because they are caused by local heating and cooling. Correspondingly, they spring up and die down rapidly. Light regional winds are most conducive to full development of the temperature differences between land and water, and between hill and valley, which give rise to local winds. Local winds have a variety of causes — often some purely local feature. We can look at some of the more common causes here.

We saw in the last chapter that land and water heat up and cool down at different rates and by different amounts through the cycle of day and night. This gives rise to the land and sea breezes common in coastal areas and along the shores of large lakes. In each case there is a return circulation at heights of 3,000–5,000 feet opposite to that at surface.

Sea breezes can be quite strong, reaching speeds of 10–15 knots. They can penetrate 100 miles inland, or more if reinforced by the anabatic effects of coastal mountains. Sea breezes can also give rise to "sea-breeze fronts." Depending on local conditions, a sea-breeze front may be indiscernible without meteorological instruments or it can be strong enough to touch off lines of thunderstorms. Sea breezes blowing from both sides of a peninsula, such as Florida, cause lifting through convergence. This is one of the mechanisms that gives Florida its frequent thunderstorms. Nocturnal land breezes tend to be weaker than diurnal sea breezes. The actual strength and direction of sea breezes is affected by the regional wind circulation, the passage of weather systems, and the season of the year. Sea breezes are affected also by the Coriolis force (see Chapter 10), so that what starts off as a breeze directly onshore in the morning may veer during the day to blow almost parallel to the shore.

Uneven heating of mountaintops and valleys gives rise to *anabatic winds* flowing into the mountains by day and *katabatic winds* flowing out of the mountains by night. The onset of these flows can be quite sudden, and, in particular, the onset of ana-batic flow up the sides of a valley can cause clouds to form suddenly along the slopes. These winds are unlikely to be stronger than 15–20 knots. Their sequence and duration are affected by various factors, among them the shape of valleys, the passage of weather systems, the regional wind circulation, the season of the year, and the amount of cloud cover. They develop most strongly in settled summer weather. Anabatic and katabatic winds are low-level flows and are unlikely to extend far above the height of the ridge crests.

Cold air pooling over a high, snowy plateau may flow down valleys as a "fall wind," "density wind," or "drainage wind." This wind is distinct from katabatic flow and is not confined to the night hours. The coasts of British Columbia, Greenland, and Norway experience such winds, as do other areas of comparable terrain. These winds can be strong or even violent; wind speeds of 100 knots have been observed in extreme conditions. A typical drainage wind might, however, rise from calm to 20–25 knots in 1–2 hours, remaining steady at that speed for several hours and filling the whole depth of the fjord down which it blows. A pilot entering such a flow from calmer, unaffected air would notice the waves on the water of the fjord in addition to other clues, not least of which would be the wind's effect on his groundspeed.

A combination of thermals rising from hot land and strong local winds can make flying conditions unpleasant on a summer afternoon. The heat will contribute to airsickness among pas-sengers. Smooth air can be found by flying higher, by flying over water if possible, or by flying earlier or later in the day.

Thermals can give rise to "dust devils"—a feature of very hot, dry climates. These can be seen as columns of swirling dust and trash, although experienced desert pilots know that they can be invisible over a paved runway. They last from a few seconds to 20 minutes and range in height from 15 feet to 3,000 feet. The stronger ones can overturn an unsecured light aircraft on the ground. Most of them can cause loss of control during takeoff or landing.

The dangerous local winds are those associated with thun-derstorms. The most extreme variety is the tornado, which is a product of violent thunderstorms. Any cumulonimbus cell can produce rapidly shifting gusty winds, and gusts of 50–60 knots are not uncommon near well-developed thunderstorms. Cumulonimbus cells produce strong vertical currents. The visual pilot unwise enough to fly beneath a thunderstorm can find himself being sucked up into it.

More treacherous to the visual pilot flying outside the storm are the downdrafts and their effects as they spread out along the ground. Downdrafts sometimes can be seen on water as "cat's paws" ruffling the surface, which either do not move or which spread out from a central point where the downdraft strikes the water. Strong downdrafts spread out ahead of the storm as a "gust front." The gust front is a layer of moving air, probably less than 1,000 feet deep, perhaps as shallow as 100–200 feet, which can be felt sometimes as far as 40 miles from the storm. Above this layer, air may be flowing in toward the storm so that a strong wind shear exists at a low altitude above ground level.

Enough has been written about thunderstorms and the effects of their associated wind shears for us to abbreviate the discussion at this point. Paradoxically, large aircraft and sailplanes are most vulnerable to wind shear and light aircraft are least vulnerable. Large aircraft are vulnerable because of their slow throttle response and high inertia. Sailplanes are vulnerable because they can regain airspeed lost because of a wind shear only by nosing down. Light aircraft are least vulnerable because of their low inertia and fast throttle response. We will look at thunderstorms some more in Chapter 7.

Winds can also be divided into *surface winds* and *winds aloft.* Surface winds are predicted in area forecasts if they are expected to be stronger than 25 knots and in terminal forecasts if expected to be stronger than 10 knots. They are reported in surface actuals and ATIS (Automatic Terminal Information Service) broadcasts whatever their strength. More on them later.

## Winds Aloft

Winds aloft are forecast for 3,000-foot height increments up to 12,000 feet above sea level and for selected flight levels at and above 18,000 feet. Winds are not forecast for any increment within 1,500 feet of the ground. The lightplane pilot considering a cross-country flight needs to know the winds at 3,000 and 6,000 feet above ground level and should know the forecast winds aloft for greater heights if he intends to cruise higher, at least up to the next forecast increment above his planned cruising altitude.

It is worth knowing where the winds-aloft reporting stations are. If the intended area of the flight includes two reporting stations, the pilot should obtain forecasts for both. If there are marked differences in wind speed and direction between stations or between successive altitudes, he should try to find out why. Turbulence strong enough to give a light aircraft a rough ride can be expected with differences in wind speed greater than 6 knots per 1,000 feet vertically and 18 knots per 150 miles horizontally. Strong wind shears aloft are common in IFR conditions but are rare in the quiet conditions conducive to good VFR weather.

In hills or mountains, the forecast wind at the elevation of the ridge crests is significant. A 20-knot wind ripping and tearing over hills will produce turbulence 2,000 feet above their crests, increasing to 5,000 feet with winds stronger than 25 knots. The smooth lift on the upwind side of a ridge is fun to ride and much sought after by sailplane pilots. Flight in the turbulent downdrafts on the lee side is more akin to sliding down a flight of steps. Wind speeds of 20–30 knots in the mountains will cause a very rough ride. If the wind at ridgecrest elevations is stronger than 30 knots, flight through the area in a light aircraft should not be considered. Even if the turbulence can be overflown, the possibility of an unplanned or partly planned descent into the turbulence cannot be ruled out. Valleys and mountain passes channel moving air, causing local increases in wind speed and, hence, turbulence.

In certain conditions of wind and atmospheric stability, a range of hills or mountains can set up a resonance in the air passing over them, with the result that a train of standing waves forms above them and persists for 100 miles or more downwind. Such waves can extend to heights as great as 40,000 feet, and sailplane pilots soaring in them have set altitude records. On passing through each wave, the air may rise and sink by several thousand feet, although the positions over the ground at which the rising and sinking take place remain constant or move only slowly. Lift and sink within the waves is smooth, although the rate of sink can exceed the ability of even quite powerful aircraft to outclimb it.

The positions of the wavecrests are marked, in suitable humidity conditions, by unmistakable lens-shaped clouds called altocumulus standing lenticularis (ACSL). If the air is moist, the waves may appear as local thickening of an overcast. If there are layers of varying humidity, lens clouds may occur in vertical or near-vertical stacks.

The formation of mountain waves depends on certain combinations of wind speed, distance between ridgecrests, distance across the mountains, wind direction, and stability of the air. Indeed, the presence of layers of air with differing stability is known to have an effect. Even small hills and islands can trigger small, isolated waves. A rare cloud formation called a pileus is formed by the same mechanism, in which stable air aloft flows over the rising crest of a strongly growing cumulus cloud.

A serious problem lies in an area of severe to extreme turbulence, called the rotor, which occurs at and below ridgecrest level downwind of the ridges causing the waves. The rotor sometimes is marked by a sausage-shaped, ragged, and visibly churning roll cloud, or sometimes not at all. Aircraft have been torn apart by rotor turbulence, and the lightplane pilot must be vigilant to avoid it. In certain conditions isolated peaks can cause stable vortices to form downwind of them, which are dangerous to aircraft in flight. These vortices may be marked by smooth, dish-like clouds.

Aviation weather services report mountain wave activity. Although sailplane pilots are much encouraged by such reports, lightplane pilots considering a flight across the affected area usually regard them as bad news.

## Surface Winds

The surface wind is often of more immediate concern than the wind aloft. In a sample of 775 accidents involving single-engine landplanes in Canada, wind was mentioned as a factor in 12% of them. The pilots lost control during takeoff or landing.

The aircraft veered off the runway onto a rough surface, hit the ground while partly airborne, hit an obstruction, or ground-looped. In a few cases aircraft were flipped over while taxiing. The pilots had begun the flight or made an approach to land without knowing that the wind conditions exceeded their ability to control the aircraft. New pilots are offered little guidance as to when they should not be flying because the wind is too strong, or as to when they could, indeed should, be flying in progressively stronger winds.

Upper and lower limits exist beyond which the situation is clear-cut. Surface winds of less than 5 knots should not trouble anyone in any type of aircraft. At the upper end of the scale, 25 knots is about the safe limit for a new pilot in an aircraft of less than 2,000 pounds gross weight, such as a Cessna 152, increasing to about 30 knots for a 3,000-pound aircraft, such as a Cessna 182. This is not to say that light aircraft do not operate successfully in wind speeds higher than these, but in high wind speeds the new pilot incurs a serious risk of an accident resulting from loss of control.

The problem is not the steady speed of the wind but, rather, the turbulence and gusts close to the ground which affect the aircraft during takeoff and landing. The aircraft's behavior in turbulence depends on its wing loading. The lighter the wing loading, the more skittish the aircraft will be in turbulence. Wing loading is a function not only of the aircraft's basic design but also of the load it happens to be carrying in flight.

The problem is twofold: turbulence and wholesale movements of the air. Most apparent is the turbulence caused by obstructions near the runway. The higher the wind speed, and the larger and closer the obstructions, the worse the turbulence will be. Many of the loss-of-control accidents attributable to wind happen at short, narrow airstrips surrounded by trees. The new pilot should make the acquaintance of such places with caution. Each small airstrip has its own quirks, and it is wise both to seek local advice and to make the first attempts at flight in calm conditions. Airports with tricky conditions are often well known by reputation, and the hazards are sometimes listed in airport directories. Larger airports, by their nature, have fewer obstructions around them. Even so, a hangar can cause turbulence for at least

1,000 feet downwind in a strong wind, and this turbulent zone may lie across the active runway in crosswind conditions.

The second problem is the wholesale upward and downward movement of the air as it clings to the ground and follows slopes. Many runways have a slope up to the threshold, and this is not always apparent when seen from the air. If a strong wind is blowing along the runway, there will be a zone of sinking air over the ground sloping up to the threshold. A pilot flying a steep approach may notice nothing. Another, flying a low, flat approach, will enter the air flowing downslope from the threshold and may need full throttle to drag himself to the runway. On occasion even that is not enough, and the baffled pilot finds himself landing short of the threshold.

Ground sloping up under the approach path has the opposite effect. The pilot unconsciously tends to maintain his accustomed height above the ground and thus flies his approach high. A wind blowing up the slope retards his descent. As a result, he lands long and may have difficulty in stopping before reaching the end of a short runway. Sloping runways and irregular terrain in the approach area cause optical illusions. A trial approach may be a good idea in some situations of this kind. The pilot should remember that he always has the option of going around for another attempt.

If the runway slopes and the wind is blowing down the slope, should we take off uphill into wind or downhill with a tailwind? Barry Schiff tackles this question in *The Proficient Pilot*. He points out that a downhill takeoff is preferable because the slope helps the aircraft to accelerate. If a downhill takeoff would mean taking off in a tailwind, the takeoff distance in that tailwind should be calculated by referring to the aircraft manual, ignoring the runway slope. If the runway is long enough, the takeoff will be safe. This method builds in a safety margin because the sloping runway will cause the aircraft to accelerate faster.

Ground sloping up to the departure end of the runway is not a problem, but hills in the departure area may be. Wooded hills and a strong wind are a bad combination. Immediately after takeoff the aircraft is flying slowly at maximum power. The hills cause turbulent downward-flowing air, which adversely affects the aircraft's climb performance. If, in such conditions, the aircraft is heavily loaded and the density altitude is high, the ingredients for a serious accident are present. Watching other aircraft taking off can sometimes be a good source of information, with the proviso that other aircraft may be more powerful or less heavily loaded than the flight being contemplated.

Because of wind effects, mountain airstrips should be carefully studied in advance. Most of these airstrips have no weather reporting facilities. Reports from larger airports in the area cannot be relied upon because mountain weather is truly local. Unpaved strips may be soft, wet, or rough and, whether paved or unpaved, they may be covered by snow and ice in winter. Telephoning the field operator about field conditions and weather is worthwhile. If a mountain airstrip has no operator and is not listed in the directories, the new pilot should question going there at all, unless he knows the area well.

If we are not very experienced, we are left wondering whether we should fly or not fly in some particular wind speed. Given that winds less than 5 knots are innocuous and winds more than 25 knots may overtax our capabilities, let us try this idea for a medium-sized airport with some obstructions nearby. Assuming that we are current on aircraft type and are reasonably well-practiced as to flying in general, we take our total flying hours and divide by 7. The result should be a safe maximum wind speed. Handling progressively stronger winds is a matter of practice and experience. Thus, a pilot with 140 hours should be able to handle a 20-knot wind safely in an aircraft of 1,500 to 2,000 pounds gross weight. Most accidents after converting to an unfamiliar type of aircraft happen within the first 12 hours on that type, so perhaps this 12 hours should be flown in light winds.

The other question concerns crosswinds. When approaching an airport, nobody has time to fool around with crosswind component diagrams. In *How to Fly: Cessna 150,* Colonel Jack Kaiser gives us this rule of thumb: When the wind is between 0° and 30° off the runway heading, the crosswind component is one-third of the wind speed. When the wind direction is between 30° and 60° off the runway, the crosswind component is two-thirds of the wind speed. When the wind is blowing at between 60° and 90° to the runway heading, the crosswind component is the wind speed. Thus, if the active runway is 18 and the reported wind is 200° at 15 knots, the crosswind component is 5 knots.

The aircraft manual lists a "demonstrated maximum crosswind velocity," which is the highest crosswind component at which the aircraft has been taken off and landed. If we cannot remember the figure for our aircraft, 10 knots is a good number. The simplest determinant of all is our ability to keep the aircraft tracking straight during the final approach to land. If we are just above the runway and are being blown sideways in spite of our best efforts to the contrary, the chances of damaging the aircraft are quite high and we had better apply full throttle and go around again. The options are to make another attempt, use another runway aligned more into wind, wait for the wind to drop, go to another airport, or, if things are really desperate, land into wind on a taxiway or on the grass. Any of these is better than damaging the aircraft in a crosswind landing.

## Finding the Wind Speed

How do we find the wind speed before takeoff, in flight, and before landing? Here are some common indicators, some of which can be seen only on the ground, others of which are visible from the air:

| Speed (knots) | Effects |
|---|---|
| Less than 1 | Smoke rises vertically. Water mirror-smooth. Absolute calm. Airport radio reports: "Winds calm." Mist and fog patches undisturbed. Windsocks hang slack. |
| 1–3 | Smoke shows direction of wind. Small wavelets on water. Slight movement of leaves. Airport radio reports: "Winds light and variable." Windsocks move. |
| 3–6 | Wind felt on face. Leaves rustle. Small waves on open water. Small twigs move. Windsocks extend halfway. Sailing boats move slowly, upright. |

6–10          Leaves and twigs move. Light flags extend. Crests on
              some waves on open water. Dry leaves picked up
              from ground. Sailing boats may be heeling but
              with little visible wake. Windsocks stream
              nearly horizontal.

10–15         Wind raises dust and loose paper. Loose paper blown
              along. Small branches move. Windsocks stream
              horizontal. Cat's paws of wind on water visible
              from the air. Sailing boats heeling and leaving
              visible wakes.

15–21         Small leafy trees sway. Whitecaps on sheltered water.
              Larger waves with white crests on open water.
              Sailing boats moving fast and heeling. Wind
              may raise dust from loose soil visible from the
              air. A Cessna 152 will rock at its tiedown. Move-
              ment of crops and trees visible from the air. Tur-
              bulence felt in flight a mile downwind of a hill.
              Light aircraft, especially taildraggers, tend to
              weather-cock into wind while taxiing. Notice-
              able ground-induced turbulence below 2,000
              feet.

21–27         Large tree branches move. Wires whistle. Small-boat
              activity curtailed. A Cessna 182 will rock at its
              tiedown. Small lightplanes may be difficult to
              control in ground-induced turbulence less than
              2,000 feet above ground level. Rough ride for
              light aircraft over hills and islands. Turbulence
              from ground up to 5,000 feet. Very rough ride
              over mountains. Difficulty in controlling small,
              light aircraft while taxiing, especially light
              taildraggers.

27–33         Walking impaired. Whole trees sway. Foam streaks
              on water. Moderate to severe turbulence within
              5,000 feet of the ground. Small light aircraft
              may be overturned on the ground while taxiing
              or if unsecured. Mountain routes unflyable in
              light aircraft because of turbulence. Crosswind
              effects perceptible when driving on exposed
              roads.

33–40         Walking difficult. Twigs break off trees. Foam in
              dense streaks on water, and breaks into spin-
              drift. Small cars may be blown sideways when

driving crosswind on exposed roads. Freshly broken small branches on ground.

40–47       Slight structural damage. Ground littered with broken branches. Some damage to power and telephone lines from tree debris.

47–55       Rare on land except in hurricanes or violent storms on exposed coasts. Power and telephone lines down. On open water, high waves and much blowing spray. Trees blown down.

56+       Widespread severe damage on land. Roofs blown off houses. Trucks blown over on exposed roads.

Purely as an aside, the south coast of Newfoundland regularly experiences winds of 80–100 knots in autumn and winter. Freight trains have been blown over, and on one occasion a wind speed of 80 knots was read on the airspeed indicator of a parked helicopter.

Even while preflighting the aircraft, we can feel the wind, whether it is rising or abating, and its gustiness, as well as observing other features of the current weather. If the airport has a Flight Service Station or broadcasts an ATIS, we can compare the present wind speed with that of the most recent hourly report. An anemometer is among the most basic of airport equipment; Unicom operators and FBOs often have one. All we have to do is walk in and ask, telephone, or transmit on the aircraft's radio.

How do we find the wind speed at our destination? Before the flight, we should, of course, obtain the most recent surface actuals for the destination and points enroute. If the flight is longer than an hour, or if conditions are changing quickly, we should try to get an update by radio on the way. If the wind at our destination is beyond our limits, we should consider what is causing the wind and what is the likelihood of its abating in the near future.

We must consider, too, the situation of the destination airport. If it is small and surrounded by trees, it might be unwise to go there even if the wind is no stronger than 10 knots.

One example of this is a small airport in British Columbia. This airport has a single 1,600-foot runway on top of a low hill. The ground slopes steeply up to each end of the runway, with approach lanes cut out of the trees. A dense woodland of tall trees bounds the airport on both sides. The surrounding hilly terrain is deceiving, and making an accurate approach to the 25-foot-wide runway is difficult. If the wind blows along the runway, an area of sinking air forms over the slope leading up to the threshold. Whichever way the wind blows, the trees produce turbulence. If the wind blows across the runway, the trees cause a turbulent layer, accompanied by a wind shear and a layer of calmer air in the sheltered zone just above the runway. One instructor with 17,000 hours remarked that he would not fly into that airport if the wind were stronger than 5 knots. The airport has a long history of crashes, some of them fatal.

Another airport on the coast of British Columbia has three paved runways on the shore of a sea strait with flat land and only low bush surrounding it. Even a 20-knot wind poses few problems there.

If our aircraft is small and lightly laden, such as a Cessna 152 flown solo, we should be considerably more cautious than if it is larger, more heavily loaded, and more powerful.

While approaching the destination, even without radio facilities, there is usually some means of discovering the speed and direction of the wind. Some of these already have been alluded to. The speed and direction of movement of cloud shadows is a guide. So, too, is the amount and direction of turbulence from an isolated hill. Small lakes and ponds show areas of calm water in the lee of the upwind shore, with waves breaking on the downwind shore. In wind speeds greater than 10 knots trees and buildings on the upwind shore will produce enough turbulence to cause "cat's paws" running downwind across the water. A bumpy flight below 2,000 feet over unobstructed terrain should alert us to wind speeds of 15–20 knots or greater. Smoke is probably the best guide of all.

One method not to use is to obtain the wind speed and direction by radio from an airport different from the destination. One pilot made the approach to a short grass runway on the basis of a wind direction transmitted to him by radio from a larger airport about 5 miles away. The wind at the grass strip was blowing in the opposite direction. Ignoring or failing to see the windsock, the pilot landed in a tailwind and ran off the end of the runway.

The positions of windsocks are marked on the airport diagrams in airport directories. We should locate the windsocks on the diagrams so that we will be able to see them from the traffic pattern. On small airfields the windsock tends to be located at midfield. On larger airports a windsock is located to the left of the touchdown area of each runway. A glance at the windsock should be one of the final checks before touchdown. If the windsock is difficult to see, chances are that it is streaming directly toward or away from us. Conversely, the stronger the crosswind component, the more visible the windsock will be when viewed from final approach.

## Turbulence

Turbulence is caused by air moving and shearing against the ground or against other layers or bodies of dissimilar air. Wind and turbulence therefore are closely related. Reviewing the amount and nature of turbulence likely to be encountered during the flight is wise. What are the likely causes? How rough will it be? Where will it be? How can it be avoided?

This is especially important when carrying passengers who are not pilots themselves. Today the average person's exposure to flight is in large aircraft flying far above most turbulence. Therefore, the turbulence normally experienced in lightplane operations, although ignored by the pilot who is accustomed to it, can be alarming to passengers. It frightens them more than any other ordinary occurrence in flight. Airsickness is as much a reaction to fear as it is to motion. Predicting and explaining any turbulence that may be encountered during the flight is well worthwhile. Then the passengers will object to it much less when it happens, and their confidence in the pilot will be much enhanced. If the flight is solely for pleasure, and turbulent conditions are expected, the flight might be arranged for another day or another time of day when the air will be smoother.

Turbulence results from the following causes:

1. *Mechanical turbulence.* This is caused by the wind blowing over hills and obstructions. Light turbulence can be expected

below 5,000 feet in winds stronger than about 15 knots and almost anywhere in mountainous areas. Moderate turbulence, which is unpleasant in a light aircraft, occurs below 5,000 feet above ground in winds of about 25 knots or greater, and especially in mountains with winds above 20 knots. Severe mechanical turbulence is encountered below 5,000 feet in winds of 30 knots or stronger and in mountain country in winds stronger than 25 knots. Turbulence in all these categories is most intense below 2,000 feet, decreasing with increasing height above the ground.

The rougher the land surface, the worse the turbulence will be. On the other hand, even a 40-knot wind can be perfectly smooth as long as nothing is there to deflect it. Mechanical turbulence can be avoided by flying high enough to overtop it, by flying over flatlands rather than hills, by flying over water rather than land. It cannot be avoided when flying specified tracks at low altitudes during departure and arrival or if clouds force the pilot to fly low.

2. *Thermal turbulence.* This is caused by hot air "bubbles" rising from heated surfaces. Notable sources are forests and rock faces exposed to the sun. Thermal turbulence is at its most severe over desert areas on summer afternoons, when moderate turbulence can extend to 15,000 feet above ground. In temperate areas the effects are much less. Thermal turbulence dies out with height. In temperate climates it is concentrated in the layer less than 3,000 feet above ground, and it usually can be overflown.

Each thermal follows a life cycle of 15–20 minutes as a bubble of heated air floats upward. The local supply of overheated air is temporarily exhausted, and the thermal dies out. Turbulence is experienced as the aircraft flies through a succession of rising thermals. Depending on the moisture content of the air, cumulus clouds may form where the thermals reach the condensation level. Smoother, but possibly subsiding, air can be found over water. Thermal turbulence begins in the middle to late morning and dies out in the late afternoon or evening.

3. *Convective turbulence.* This is associated with large cumulus clouds, towering cumulus, or cumulonimbus. Such clouds contain strong updrafts whose development is followed by

equally strong downdrafts. The shearing between them can cause severe turbulence. The more strongly developed the cell, the more violent the turbulence is likely to be. Violent thunderstorms can generate turbulence strong enough to destroy an aircraft in flight, even in the clear air outside the storm cloud itself. Hence the rule of thumb: Any thunderstorm should be avoided by 5–10 miles, and a severe one by 20 miles.

When a cumulonimbus cell is moving, the most turbulence will be found ahead of it, the least behind it. In varied terrain, convection is strongest over hills, weaker over flatland, and weakest over water, unless the warmth of the water relative to the land is itself the cause of instability. Flight between and around well-developed cumulus clouds, perhaps a mile across and a mile deep, might encounter fairly sharp jolts in the surrounding clear air. Moderate turbulence can be expected within 1–2 miles of towering cumulus. Cumulonimbus cells, whether they have or have not developed into thunderstorms, should be avoided by wider margins.

4. *Wind shear.* If air movement differs with height, a layer of turbulence occurs where one layer of air shears over another. Weak wind shears aloft are common on calm evenings; a climb or descent will find smooth air. Strong wind shears are associated with thunderstorms and fronts and are common in IFR conditions.

5. *Fronts.* Fronts are a fruitful source of turbulence of all kinds, but such turbulence is more likely to be encountered on IFR than on VFR flights.

6. *Wake turbulence.* When approaching to land behind another aircraft (all of which leave some degree of wake turbulence), the pilot would be considerate to remark to passengers on the possibility of wake turbulence. Any aircraft larger than a light twin produces wake turbulence strong enough to threaten the control of a small lightplane during final approach and landing, especially during the landing flare. The wingtip and flap vortices trailing behind large jet aircraft are especially dangerous in this

respect. Their avoidance is dealt with sufficiently in basic training to warrant no further mention here.

Wind and the motions of the air have a stronger and more limiting effect on the operations of a light aircraft than on those of larger and heavier ones. The new pilot has good reason to make a careful study of the wind.

# 4

# Day, Night, Seasons, and Terrain: Performance Effects

## Do aircraft fly better in the dark?

HAVING DISCUSSED various sky conditions affecting the aircraft's performance, which change rapidly and to some extent unpredictably, it is comforting to know that some conditions change either regularly and predictably or not at all.

The cycle of day and night are absolutely predictable. The seasons vary in intensity and duration year by year but otherwise are permanent features of our lives. The terrain itself is, for all practical purposes, absolutely unchanging. All of these factors affect the atmosphere in ways that in turn influence the light-plane's performance.

## Effects of Day and Night

As far as this chapter is concerned, the effect of day and night is a cyclic fluctuation in temperature and all that that entails. We saw in Chapter 2 that the earth's surface receives energy from the sun and radiates some of this energy back into space.

The rate of energy flow back and forth is influenced by the nature of the surface — water, soil, rock, forest, snow — and by the thickness of cloud cover.

At night, when solar radiation is absent, radiation from the earth causes its surface to become cooler, and this cools the air closest to it. Other things being equal, the air within a few thousand feet of the surface becomes cooler during the night. Things that may not be equal are the movement of pressure systems and fronts.

As soon as the sun rises, the earth begins to receive solar radiation and to become warmer. As a result, the bottom part of the atmosphere also becomes warmer. The intensity of energy inflow from the sun depends on the sun's angle in the sky and the amount of cloud cover. Consequently the diurnal air temperature range may vary between a few degrees and extremes of 55°C/100°F.

We saw in Chapter 2 that the air temperature affects the aircraft's performance. We also learned how the independent variables of surface elevation and temperature are combined in a concept called density altitude. Of these two variables, temperature goes through a cyclic change during the 24 hours. Therefore, any place on or near the ground goes through a cyclic variation in density altitude. Density altitude is at its lowest at daybreak and at its highest in the middle of the afternoon.

This fact is of critical importance in summer at the many high-altitude airports throughout the western parts of North America. Some of these airports are one-way strips because of a sloping runway or because of hills at one end of the field. A tailwind in the takeoff direction can further complicate matters. On clear summer mornings these places heat up quickly, accompanied by a correspondingly rapid increase in density altitude. Thus, prompt departure in the morning may be mandatory.

As mentioned in the previous chapter, local winds result from temperature differences between land and water, mountain and valley. These differences are at their greatest just before sunrise when differential cooling is at its maximum, and in the middle of the afternoon when differential heating is at its peak. These effects are at their least in the morning and evening. Therefore, winds induced by thermal contrasts and the mechanical turbulence associated with them are at a maximum in the early morning and in midafternoon. Temperature contrasts caused by differential radiation at night are rarely as marked as those caused by the differential absorption of energy during the day. Therefore, local winds tend to be weaker at night than during the day. This is not necessarily true of strong local winds caused by convective activity. We also have noted the occurrence of low-level jetstreams, a nocturnal phenomenon.

Thermal turbulence induced by local heating of land surfaces is at a maximum during the afternoon and minimal or absent in temperate latitudes at other times of the day.

We will see in Chapter 9 that atmospheric stability goes through a daily cycle. The most important effects of stability are on clouds, precipitation, and the pilot's ability to see, but we should note here that unstable conditions can produce wind and turbulence in great abundance. The lower atmosphere tends to be more stable during the night and less stable during the day, for reasons discussed in Chapter 9.

Night-time is especially conducive to the cool, damp conditions that favor the formation of carburetor ice. Heating during the day increases the air's capacity for water vapor. When the air cools in the evening, the spread between temperature and dewpoint closes and the air may approach saturation. The smaller the temperature-dewpoint spread, the more likely carburetor ice is to form.

Night is when frost forms on wings, and it must be removed before takeoff. Night-time, especially on weekends, may yield little in the way of heated hangars or antifreeze. The cunning pilot flying on a winter night therefore will either take antifreeze with him or will bear in mind that fuel from the sump drains will remove frost.

## Seasonal Effects

Each season of the year brings phenomena that improve or degrade the aircraft's performance. Density altitudes, which can cause performance problems in summer, are much lower in the lower temperatures of winter. Mountain airstrips, however, now become soft and wet or are covered by snow and ice. This problem is not confined to mountain airstrips although, as a group, they tend to be the most isolated and least maintained. Any small airstrip, be it the departure or the destination airfield, should be checked in advance if waterlogging or snow cover is likely to be a problem. Takeoff performance is so degraded by snow or soft, wet turf that takeoff may be impossible in the available distance. Icy runways and snow-covered taxiways and ramp areas are also a feature of the winter months; experienced advice should be sought on how best to deal with the hazards they present.

If generalization is valid at all, regional winds caused by circulation around pressure systems are strongest in winter and weakest in summer. Local winds produced by thermal contrasts are stronger in summer, with the exception of winter "drainage winds," which can be quite strong. Thunderstorms and tornadoes occur most frequently during the summer months. In affected areas certain periods of the year are known as "tornado season" or "hurricane season." With the exception of winds generated by convective cells, local winds seldom reach the strength or persistence of regional winds. Therefore, in general, the periods when lightplane flying is most impeded by wind are longest and most frequent in winter.

## Effects of Terrain

The terrain over which we fly can be anything from flat lowlands at sea level to jagged mountains towering above the service ceilings of some aircraft. The effects of high and rugged terrain are too many and varied to be covered in this book. A whole book, *Mountain Flying,* by S. Imeson, is devoted to the subject.

Dual instruction should be sought in the techniques to be cultivated and the pitfalls to be avoided when flying in the mountains. Every year flatland pilots fly into the mountains and crash. Sometimes the aircraft and its occupants are lost without a trace; sometimes the occupants escape or are rescued after harrowing experiences. Many of the pitfalls that cause these accidents stem from the effects of terrain elevation and topography on aircraft performance.

Mechanical turbulence is a function of wind speed and terrain roughness. Surface wind speeds over open water are higher than over land, generally by about 10 knots, because of the lower frictional resistance, but turbulence is less for the same reason. The rougher the terrain, the more severe is the turbulence generated by a wind of any given strength and the higher above the terrain crests the turbulence extends.

The more rugged the terrain, the stronger and more frequent are upward and downward movements of the air. In mountain country, downdrafts can easily exceed a light aircraft's best rate of climb, especially if it is already degraded by the effects of altitude or temperature. Sometimes it is supposed that a downdraft must spread out when it strikes the ground and for that reason an aircraft cannot be thrown to the ground. Unfortunately, this is not true. Air flowing against the windshield of a car changes its direction; a bug flying in that air cannot do so, because of its inertia, and bursts against the windshield. A downdraft can force an aircraft to the ground for the same reason, or it can put the aircraft into a position from which it cannot escape collision with an obstacle. Most dangerous of all mountain turbulence is the rotor turbulence associated with mountain waves.

By following the lay of the land, rather than plowing ahead in a straight line, the cunning pilot can exploit rising air and avoid sinking air. Smooth and often powerful lift can be found on the windward sides of ridges. Sailplane pilots have used this lift for decades. Lightplane pilots can avail themselves of it in the same way, although they should keep a sharp lookout for sailplanes, ultralights, and hang gliders doing the same thing. Correspondingly the leeward sides of ridges are areas of turbulent sinking air and should be avoided. Lift is found along sunlit slopes, and sink along shaded ones.

In most conditions of convective turbulence, convection is more intense over hills than over flatland, and more intense over land than over water. The exception is when the warmth of the water relative to the land is the cause of the convection. The sun's heating of land surfaces touches off convection while the air over the cooler water remains unaffected. Correspondingly, hills provide a measure of lifting which is absent over flatland. Even slight movement of the air around the hills is enough to provide significant lifting, with the result that we see puffy cumulus clouds floating over hills in apparently calm conditions. The air rising in convection currents may spread out at heights of 5,000–10,000 feet and then descend, with the result that those areas free of convective clouds may also be areas of sinking air, requiring higher throttle settings to maintain airspeed and altitude.

Local terrain effects are fascinating in themselves, and new pilots should miss no opportunity to build up a store of knowledge about them.

Day and night, the seasons and terrain all affect the aircraft's performance. These effects are predictable. New pilots must learn to predict them, adapt to them, allow for those that degrade the aircraft's performance, and turn the beneficial ones to their advantage.

# 5 Disorientation
## What happens when you don't know which way is up?

**DISORIENTATION KILLS PILOTS,** especially new ones. It results from conditions of weather or darkness, or both, which hide the earth's surface. It is therefore well worthwhile to look into how it occurs, what effects it has, and the conditions in which it can strike.

Once we are airborne, the bodily senses, other than sight, which we use to remain upright or to detect changes in our attitude relative to the earth's surface become useless. We must have a visual reference to show us which way is up, and we must believe this reference rather than our bodily senses. This can easily be demonstrated by flying with another pilot and trying to maintain control of the aircraft with our eyes shut. Even birds cannot fly when blindfolded. When reference is provided by being able to see the ground, it is easy to stay upright. When sight of the ground is cut off, we must rely on the aircraft's instruments. The skill of flying by instruments can be acquired only through rigorous training, and retained only through constant practice.

## Causes of Disorientation

When we are on the ground, our primary attitude reference is the surface of the earth, supplemented by the effects of that force field known as gravity. By seeing the ground and by feeling the loads gravity imposes on us, we know which way is up. We also have little channels inside our ears that can detect accelerations and changes in attitude and direction by means of the inertia of a fluid inside them. These reference systems work as long as we are moving about under our own power. They even work quite well when we cannot see, as we know from the children's game of blindman's buff.

The trouble starts when we are moved about in a vehicle. To experience this, we need only to have someone drive us around town while we have our eyes shut. Faced with a series of decisions based on inadequate information, we become progressively disoriented. Each decision forms the basis of another, and quite soon the web of wrong decisions is impossible to unravel. Panic accelerates and completes the disorientation process.

When we are on the ground, at least we know which way is up, even though we may be blindfolded or enclosed in a vehicle. In flight this is no longer true. An aircraft in flight subjects its occupants to accelerations they do not experience on the ground, through turbulence, maneuvers, or both. We are accustomed to having the force of gravity acting straight downward toward the earth, but accelerations of the aircraft can produce other forces that we perceive falsely as being that of gravity. Although the force may point downward relative to the aircraft, it does not necessarily point toward the earth's surface. As a result, if we cannot see the ground or the top of a cloud layer that probably approximates it, we do not know where the earth's surface is. We may feel a pseudo-gravity produced by the aircraft—and even this can change quickly—but the aircraft itself remains under the influence of the real gravity which we can no longer perceive.

This can be illustrated by the forces acting on the pilot in a loop. The force exerted by the pilot on his seat in upright, straight, level flight results from the force of gravity alone and is equal to 1 g. Let us say that the centrifugal force caused by the aircraft's change of direction during the loop is 2 g. As the pilot starts the loop, the two forces are pointing in the same direction—downward—and he feels 3 g pushing him into his seat. As he comes level inverted at the top of the loop, the centrifugal force generated in the looping aircraft is still 2 g but now is pointing straight up—away from the ground. The force of gravity is still there—1 g acting downward—so that the net force acting on the pilot is 1 g straight up. If the pilot closes his eyes, while seated firmly in his seat under the influence of 1 g, his remaining senses detect nothing that would tell him that he is upside down.

Depending on the aircraft's maneuvers, the pilot can be in positive g, weightless, or in negative g in any attitude of the aircraft relative to the earth. The aircraft produces acceleration forces, which may be added to or subtracted from the force of gravity and which are indistinguishable from it, yet which operate in directions with no fixed relationship to the earth's surface. That is why aerobatics at night or in reduced visibility can be extremely dangerous.

The body can be tilted to appreciable angles, as much as 20°, from its upright position without the person's being aware of it if the tilting is done slowly enough. The body can also be turned slowly around with a similar lack of awareness. If the tilting or turning stops abruptly, the person, unaware that it was going on in the first place, believes that a tilting or turning motion has begun in the opposite direction. Fore-and-aft accelerations produce illusions of pitching up or down, and there are others. To describe the illusions that affect pilots flying in instrument conditions is not necessary here, except to say that they are numerous and convincing.

Pilots may even develop their own illusions. One pilot, flying a high-winged aircraft in instrument conditions, caught glimpses of the ground almost directly below the aircraft on its lefthand side and felt a powerful sensation that the aircraft was in a steep bank to the right. Other instrument pilots denied having experienced this particular illusion.

## Effects of Disorientation

Pilots without full instrument training and recent practice who find themselves deprived of outside visual references will succumb to the effects of these illusions in a short time. They may be unaware that the aircraft is in an undesirable attitude and therefore fail to correct it. Alternatively, they may perceive the aircraft to be in an unwanted attitude even though it is flying straight and level, and therefore may "correct" it into the unwanted attitude they are trying to avoid.

Even if the pilot does not know what is going on, the aircraft is still under the control of gravity and its own aerodynamics. These soon seize control from the pilot, and one of three maneuvers commonly results. Tests have shown that visual pilots deprived of outside visual reference can sometimes last as long as 3 minutes before losing control, but most do not. These maneuvers are the roller coaster, the spin, and the spiral dive.

An aircraft trimmed in a climb, in level flight, or in a descent at a constant power setting is stable in the pitching plane. If turbulence or a tweak on the yoke should cause it to pitch up or down it will, unless disturbed again, return to its trimmed attitude in a slow, porpoising motion called a *phugoid*. The period of a Cessna 172's natural phugoid is about 15–20 seconds, in which time the pilot or further turbulence may interfere. The instrument-naive pilot who enters clouds may see and hear the airspeed increasing or decreasing and cannot let well alone. He starts to chase the airspeed needle, which leads to wilder and wilder excursions from the trimmed flight attitude. Various things can happen even if the aircraft continues to fly straight, which it may not.

The roller coaster consists of a series of steeper and steeper climbs and dives. The aircraft may emerge from the cloud base in a steep dive. In that case, it may hit the ground before the pilot can recover; the pilot may pull the wings off in his attempt to pull out of the dive; or the aircraft may stall as the pilot attempts to pull out and then spin—not necessarily an upright spin either. The other possibility is that the aircraft, zooming upward in clouds, ends up in a vertical climb or inverted before whipstalling and going into an upright or inverted spin.

Aircraft in straight and level flight do not spin all of a sudden, either in clouds or out of them. What happens in clouds is that the pilot, having no attitude reference, gets the aircraft into a steeper nose-high attitude than it can sustain. The result is a power-on stall, which is more violent than a power-off stall and usually involves a snap off onto one wing. Without immediate and correct action by the pilot, the aircraft then spins. Recovery from a power-on stall demands prompt and correct action. This is difficult enough in clear air with ample horizon references. In clouds it is unlikely to be carried out correctly, if at all.

A power-on spin is much more violent than the power-off version taught in basic training under carefully controlled conditions and may be irrecoverable as long as power is left on. Indeed the power-on, high-speed, high-g maneuvers of an aircraft out of control cannot be taught or practiced for fear of overstressing the airframe. Power-on spins, tailslides, and whipstalls are forbidden maneuvers for all but a few specially built aerobatic aircraft, and even then are regarded as advanced maneuvers to be flown only by skilled aerobatic pilots. It should be obvious that a pilot involved in these maneuvers blind in clouds in a normal category aircraft is unlikely to survive. Aircraft not uncommonly break up in flight after control is lost in blind conditions.

Unless an aircraft is perfectly balanced laterally, it will, if left to its own devices, begin a slow spiral, turning toward the heavier side. This maneuver is self-reinforcing. Without external references the pilot may be unaware that the aircraft is turning and may believe that it is going down faster and faster in a straight dive. He hauls back on the yoke, which merely tightens the spiral. The wings may come off, or he may emerge from clouds too late to avoid hitting the ground. Spins may degenerate into spiral dives or the disoriented pilot may recover from a spin in one direction to a spin in the opposite direction.

The result of any of these maneuvers is that the aircraft is heading toward the ground at high speed out of control—assuming that it remains intact. Any pilot, no matter how skillful, would need something like 1,000 to 2,000 feet to recover a light aircraft from such a situation. Heavier aircraft require more height. The visual pilot who enters clouds inadvertently almost certainly does so at low altitude. Typically he will have been flying lower and lower in worsening visibility, with the cloud and the ground converging, precipitation increasing, or night beginning to fall. He will lose control in clouds, fog, precipitation, darkness, or a combination of these, either because he lost outside horizon reference in level flight or because he tried to climb through the cloud on instruments. Having lost control, he will be unlikely to regain it before the aircraft breaks up in flight or hits the ground.

## Conditions Causing Disorientation

The visual pilot does not have to be in thick clouds or fog to lose outside horizon reference. By day in rain, snow, or misty conditions, unbroken areas of calm water or smooth fields of snow or ice may not provide sufficient reference. Certain light conditions between a snow-covered surface and a low overcast, with or without snow falling, can produce a condition known as whiteout, in which all horizon references vanish. When faced by whiteout, polar explorers on the ground give up and pitch camp. With the onset of darkness, these conditions become more acute.

If a pilot wanted to fly "VFR on top" between cloud decks, he would find that a good horizon reference by day would fade out completely by night.

The "horizon" is assumed to be horizontal. In mountainous country this is often untrue. Flying low down amongst high terrain where nothing is horizontal is conducive to disorientation. Any obscuration, such as mist, rain, snow, or darkness, that blurs the available attitude references is especially dangerous in mountainous country.

Night is the great obscurer of visibility. Horizon references can disappear over uninhabited country or open water in the absence of light from stars or the moon. Clouds can be flown into before they are seen. Fog and stratus tend to form at night, which adds to the problem. Night illusions are a subject in themselves and are discussed in Chapter 11.

To summarize this chapter, we should say that the senses, other than sight, for telling us which way is up do not function reliably in an aircraft. Visual pilots depend on being able to see a horizon reference outside the aircraft or they will lose control. Once out of control at low altitude, the aircraft is almost certain to break up in flight or hit the ground at high speed before control can be regained. Many conditions of intermediate visibility besides actual immersion in clouds can obscure the necessary horizon references. Any condition of the sky that threatens to do this is extremely dangerous to the new visual pilot and must be avoided.

So dangerous is loss of horizon, in fact, that the threat of it should be treated as a full-blown emergency requiring a precautionary landing unless the pilot is fully capable and equipped to continue the flight by instruments. If a suitable landing ground is available, and if the landing is carried out properly, the risk of damaging the aircraft is relatively slight. The risk while landing under control is in any case negligible compared to the much greater probability of destroying the aircraft and killing or injuring the occupants by becoming disoriented or by flying at full speed into an unseen obstacle.

In the next chapter we will see the effects of ceiling and visibility on the visual pilot's flight.

# 6

# Ceiling and Visibility

## How high is the sky?

**SOME DAYS** the fishbowl of visibility in which the visual pilot flies is infinitely large—mountains can be seen a hundred miles away and the pilot can look up and out into space. Other days the margins of that fishbowl are alarmingly constricted—a gray cloud ceiling crushes pilots down toward the ground, and they fly in a constantly moving circle of visibility, where ground features appear ahead and disappear behind. We have here the two very important issues of ceiling and visibility.

## The Visual Flight Rules

The legal minima for flight under the Visual Flight Rules differ slightly between Canada and the United States.

In the U.S. aircraft must be flown 500 feet or higher above obstructions, except over water or in uninhabited country, where no legal minimum altitude is specified. When flying VFR at less than 1,200 feet above the ground, the aircraft must be kept clear of clouds and forward visibility must be at least 1 statute mile. In controlled airspace this requirement increases to 3 statute miles. If clouds are more than 1,200 feet above the ground, the aircraft must be kept not less than 500 feet below them, 2,000 feet horizontally from them, or 1,000 feet above them. When the aircraft is flying 10,000 feet or more above sea level, these requirements increase to 5 miles visibility, 1,000 feet below, 1 mile horizontally, or 1,000 feet above clouds, except where the terrain is so high that the flight is less than 1,000 feet above the ground.

In Canada the regulations concerning minimum flight altitudes are similar in substance to the U.S. regulations. In uncontrolled airspace, when less than 700 feet above ground level, the aircraft must be clear of clouds; visibility must be at least 1 mile (2 miles in coastal British Columbia). More than 700 feet above ground level, the required flight visibility is the same but the aircraft must be 500 feet or more vertically and 2,000 feet or more horizontally from clouds. In controlled airspace the flight visibility must be at least 3 miles and the aircraft must be 1 mile or more horizontally and 500 feet or more vertically from clouds. "VFR on top" is not permitted in Canada, where VFR flight must be conducted in sight of the earth's surface.

In aviation weather service area and terminal forecasts VFR conditions are defined as ceilings more than 3,000 feet above ground level and visibilities more than 5 miles. IFR conditions are defined as ceilings less than 1,000 feet above ground level and visibilities less than 3 miles. Marginal VFR includes conditions between the two.

## Ceiling

Now that we have said all that, it is essential for the new pilot to realize that the legal minima for VFR are no guarantee of safe visual flying conditions. We can fly at low level over empty country at night beneath a low, thick overcast in visibility such that we could not even see across a decent-sized airfield in broad daylight and still be flying legally under the Visual Flight Rules. How long we would survive in such conditions is conjectural. Because the legal definition of VFR offers no safe guidance, let us take a look at the ceiling and visibility conditions that we actually do need for the safe conduct of our flight.

Convenient cruising altitudes for the lightplane pilot are somewhere between 2,500 feet above ground level and 10,000 feet above sea level. Viewed from a substantial height, the land is laid out like a chart, and widespread combinations of landmarks are clearly visible. The apparent speed over the ground is slow enough for the land to be inspected and for the chart to be read at leisure. The aircraft can receive navigation and communication radio signals from great distances and, merely by listening, the pilot often can be warned of bad weather or other problems far in advance. The aircraft is likely to be on someone's radar screen. In the event of an in-flight emergency, especially one involving loss of power, time and space are available to rectify the problem if possible and, if not, to glide down to a suitable forced-landing ground or possibly an airport.

Then the clouds come down on top of us. In the U.S. we can climb above them, which leaves us with the problems of navigation and how to get back down again. In Canada we are not allowed to fly above anything denser than a scattered layer. We know that we cannot fly through the clouds themselves; therefore, unless we can climb up over them or weave between them, we must permit ourselves to be forced down toward the ground. This is no problem down to 3,000 feet above ground level, but then the difficulties begin.

From 3,000 feet down to 1,000 feet, as the clouds crush down above us, we begin to lose contact with radio navigation and communication facilities, especially in hilly country. Landmarks appear fewer at a time and in sharper perspective, which tends to alter their apparent shape. They disappear behind us that much sooner. Our choice of forced-landing grounds diminishes. Flight is now somewhat inconvenient, no matter how good the visibility may be below the broken or overcast cloud layer. With an overcast at that height, visibility is seldom perfect, if only because it is darker under the clouds. Night will come sooner. Now, too, unless the land is perfectly flat, the hills and undulations decrease still further the distance between the clouds and the ground and begin to hide landmarks off to one side of our course. Our route may have to start snaking along valleys. Our flight is inconvenienced — greatly inconvenienced if we get lost! — but not hazardous.

Below 1,000 feet — and remember that we are not flying below 1,000 feet for the fun of it — the flight becomes hazardous. One thousand feet is no more than the altitude of the traffic pattern around most airfields. Some transmission towers are a thousand feet high; some of them are on hills and ridges, possibly with their tops in the clouds and their guy wires spread out for some distance around them. Navigation now becomes very difficult. Unless the area is densely populated, we may be cut off from radio voice communication; we may or may not appear on radar; we very likely will be unable to receive navaid signals unless we are close to the transmitter. Even a wooded hill now cuts off much of our available airspace. Below 500 feet manmade obstructions — bridges, towers, tall buildings, high-tension cables, chimneys — multiply rapidly. Unless we are over the open prairies, in which case the risk of becoming lost is even greater, this bottom 500 feet of the sky is a dangerous place to be. If the ceiling is this low, visibility is unlikely to be at all good. Low ceilings are often ragged as well, and the extent of the raggedness can be several hundred feet. Below 1,000 feet, in the event of power failure, the pilot may just have time to turn into wind; his selection of forced-landing grounds will be very limited.

Taking everything into account, if the ceiling is between 3,000 and 1,000 feet above ground level, navigation is difficult except in familiar territory, and the risk of becoming lost is correspondingly great. Under a 1,000-foot ceiling, flight may become dangerous; under a 500-foot ceiling, it is very dangerous indeed. Over water, at dusk, at night, or in hilly or mountainous terrain, these dangers are vastly increased.

Flight through certain mountain passes is not recommended unless the ceiling at stations near the entrance to the pass is 5,000 feet or even higher. A ceiling of 2,000 feet may be acceptable in some wide valleys and intermontane basins, but the bottom 2,000 feet of a winding, V-shaped canyon is not a safe place to be. Mountain weather is local and can change quickly. For these reasons the new pilot would be well advised to fly in the mountains only in fine weather in daylight.

## Visibility

In open, flat country we will notice some sort of obscuration when the visibility is down to 15 miles. As long as we are competent navigators and are flying reasonably high, this will not be a problem. The same remains true in visibilities down to about 5 miles. If we can follow highways, railroads, or shorelines for a good portion of the route, 5-mile visibility should pose no problem. When the visibility is 5 miles, we will be aware that only a limited selection of landmarks is presented for our inspection at any one time. Large sectors of our field of vision are blocked by the aircraft itself. More distant landmarks, which otherwise would be visible beyond the aircraft's nose or wings (if low wings) or fuselage, are now lost in the haze, mist, or rain. A landmark may swim into our field of vision and then fade out before we have had time to fit it onto the map. We identify things more than we realize by means of patterns and combinations rather than by single features in isolation. We are therefore at a disadvantage if we cannot see an assemblage of landmarks. Accurate dead reckoning and electronic navigation come into play when we are in 5-mile visibility in unfamiliar territory.

These conditions persist down to visibilities of 2–3 miles. Then the problems begin. With visibility of 5 miles or better, we can see areas of worse obscuration ahead. In visibilities of less than 2 or 3 miles, a vicious circle comes into play. The shorter the distance we can see ahead, the less easily we can see areas of diminished visibility and the more serious are the effects of such reduced visibility. In visibilities of 1 to 3 miles, all may be well over familiar flat terrain, although our downward slant visibility will be a cone of uncomfortably small extent. If we could be sure that the visibility ahead was nowhere less than 3 miles, we could cruise on in reasonable peace of mind. In haze this may well be the case, but in rain or mist there is no guarantee that it will not become worse. In fact, it has only to become a little worse and we are in serious trouble.

Let us think about visibility down around a mile. That means we can only just see the far end of a 5,000-foot runway when we are about to land on it. We can barely see the buildings

on the far side of an international airport. That is not much
visibility to be flying about in. Such conditions often are associated with stratus-type clouds whose edges are indeterminate,
ranging from thin mist to thick cloud. Without direct experience
it is difficult to conceive how nearly impossible it is to see one
patch of cloud against another in reduced visibility and poor
light. Such conditions are a deathtrap, and this is how it springs
shut.

We are flying between 500 and 1,000 feet above ground in
visibility of 1–1½ miles in mist and light rain. The horizon ahead
is indeterminate. Sometimes we see a little more, sometimes a
little less. Sometimes rags of cloud drift past not far away. We
know that the visibility is restricted, so the absence of a clearly
visible horizon ahead does not surprise or concern us. It should!
We look down at the ground to one side because that is our only
attitude reference. We cruise on for a while. Conditions have been
getting steadily worse, and we probably are beginning to wonder
where we are. Our sense of time and space has deserted us. Instantly the ground is rubbed out because we just flew into cloud.
What happens next is anyone's guess.

This sequence of events happens not infrequently even to
student pilots flying in the airport traffic circuit, where supposedly the sky condition is well known. Mostly they duck down
and come out into the clear; sometimes they lose control with
fatal results. If it happens as easily as that over an airport, it
should be obvious that this type of situation is extremely dangerous on a cross-country flight, especially with night coming on.

Often, when flying in controlled airspace, a radar air traffic
controller assigns an altitude and a heading. Because the controller cannot see clouds on his radar scope, the need for the visual
pilot to maintain outside horizon reference takes precedence over
any assigned altitude or heading. The words, spoken over the
radio, "unable to maintain VFR" are sufficient reason for flying
any altitude or heading that will "maintain VFR," provided the
pilot makes the controller aware of what is going on. If the
weather is bad and the traffic dense, the controller has the option
to refuse to let an aircraft into his territory or to tell the pilot to
leave if the aircraft is already inside it. That is his prerogative.

Our lives depend on "maintaining VFR." The rumored cause

of one accident was that the pilot was so spellbound by the controller's vectors and altitudes that he flew into a cloud and lost control, with fatal results. Another pilot was killed on a night cross-country flight because the controller told him that airport conditions were below VFR minima even though the pilot could at one time see the runway. Having been refused landing permission, the pilot flew away into the darkness and foul weather. If things are bad enough, we can declare an emergency and do anything we want within reason—and then argue about it afterward.

We are accustomed to flying several thousand feet above the ground. When we fly over familiar territory at 1,000 feet, all sorts of obstructions loom up which we did not notice before. The hills we always fly over now tower above us. The same is true of radio masts and powerlines strung across rivers. But that is not the whole story. The most usual cause of poor visibility is saturation of the air with moisture. In such conditions the balance of temperature, pressure, and humidity is delicate. Very little is needed to cause the moisture in the air to condense. When this wet air flows over a hill, between hills, or down a valley, there is a venturi effect which can easily cause local condensation—clouds. Thus, when we are flying low in poor visibility not only does the local terrain loom up larger than we expect, but that terrain also can hide in clouds of its own making.

One often-heard statement that can be disastrously untrue is: "It's always better when you're up there." Once an aircraft is airborne, the view can sometimes expand, and obscurations such as rain are not as critical as was feared on the ground. Accidents have happened, however, when pilots—some of them instructors—have taken off to "test the weather" and have flown at once into conditions that deprived them of ground references. This is easily done at night, in falling snow, or where patchy stratus hides in falling rain against a stratus background.

Such are the difficulties and dangers that reduced ceiling and visibility can cause to the lightplane pilot's visual flight. Shown here are some maps (Figures 6.1–6.4) indicating the percentage of the time when ceilings are below 1,000 feet and visibilities less than 3 miles. These may be conditions that are legally VFR in

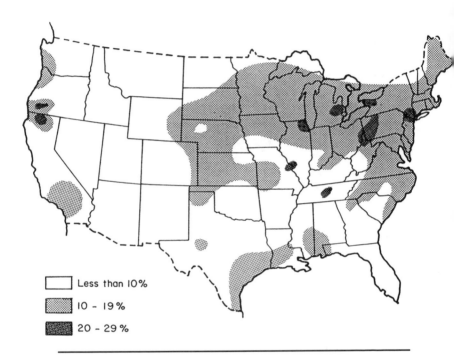

Less than 10%

10 - 19%

20 - 29%

Fig. 6.1. Percentage of time when ceiling is below 1,000 feet AGL and visibility is less than 3 miles: Spring. Adapted from *U.S. Air Force Weather Manual* (AFM 51–12).

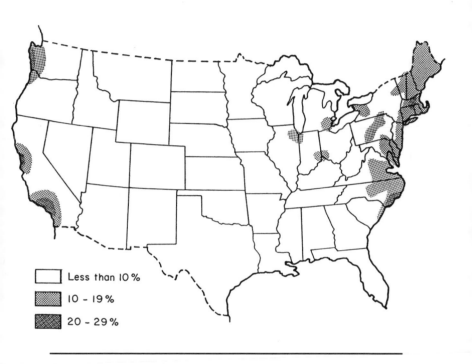

Less than 10 %

10 - 19 %

20 - 29 %

Fig. 6.2. Percentage of time when ceiling is below 1,000 feet AGL and visibility is less than 3 miles: Summer. Adapted from *U.S. Air Force Weather Manual* (AFM 51–12).

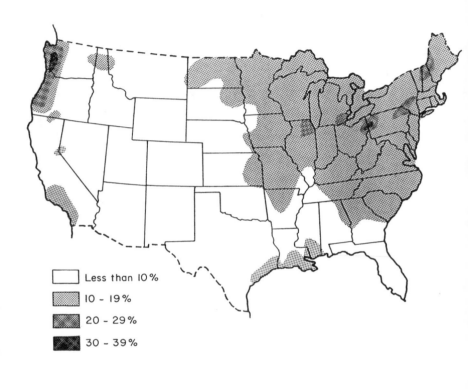

Fig. 6.3. Percentage of time when ceiling is below 1,000 feet AGL and visibility is less than 3 miles: Fall. Adapted from *U.S. Air Force Weather Manual* (AFM 51–12).

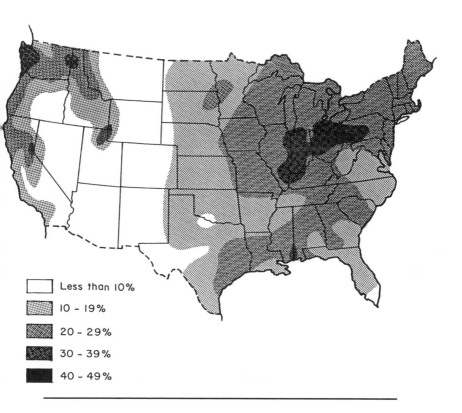

Less than 10%

10 - 19%

20 - 29%

30 - 39%

40 - 49%

Fig. 6.4. Percentage of time when ceiling is below 1,000 feet AGL and visibility is less than 3 miles: Winter. Adapted from *U.S. Air Force Weather Manual* (AFM 51–12).

some areas, but new visual pilots cannot fly in them on cross-country flights with any degree of safety. These are the percentages of the time when the visual pilot should have to cancel his intended cross-country flight, especially if night flying is involved as well. The pilot who "always gets through," yet who operates in an area with IFR conditions 30% or more of the time, could be headed for an accident.

Ceiling and visibility are sky conditions which the new pilot should consider as if his life depended on them. It does. In the following chapters we will look at features of the weather that affect ceiling and visibility and how they are likely to do so.

# 7

# Water in the Atmosphere

## In unsuspected forms

**ONE OF THE CONSTITUENTS** of the atmosphere is water. Even though it makes up less than 4% of the air by volume, its effects are disproportionately great. Except when it turns to ice, water seldom affects the performance of a light aircraft to any significant degree. It does, however, cause opacity of the air, which has profound effects on our ability to fly.

Water can exist in the atmosphere in all three states — solid, liquid, and gas. We know ice, water, and steam (at least we think we do), but in the atmosphere things are different. We think of steam and water vapor as one and the same thing, which is inaccurate. In particular, we think of steam as the billowing white clouds we can see. In fact, *water vapor* is an invisible gas and *clouds of steam* are clouds of water droplets that have condensed from the vapor. If they are droplets of water, why do they not fall?

We are also familiar with the idea that water exists as a solid (ice) at temperatures below 0°C/32°F, as a liquid between 0°C/32°F and 100°C/212°F, and as a gas above 100°C/212°F. In the atmosphere, however, water can exist as a vapor at all temperatures found on earth. It can exist as a liquid down to $-40°C/$ $-40°F$; it can exist as ice in air temperatures slightly above freezing. We can see various ideas emerging to which we are as yet strangers.

## Saturation and Condensation

The amount of invisible water vapor the air can hold depends on its temperature. The difference between what it is holding at some particular temperature, time, and place, and what it could be holding at that temperature is called the *relative humidity,* expressed as a percentage. The mass of water vapor per unit mass of dry air is called the *mixing ratio.* The mixing ratio that will cause saturation (the *saturation mixing ratio*) is not by any means constant but increases sharply with increasing temperature. In other words hot air will hold far more water vapor than cold air. The saturation mixing ratio doubles for every $11°C/20°F$ increase in temperature. At $-15°C/+5°F$, 1 kilogram (2.2 pounds) of dry air holds 1 gram (0.03 ounces) of water vapor at saturation. At $+35°C/95°F$ that kilogram of dry air holds 40 grams (1.40 ounces) of water vapor at saturation. This explanation is neither profound nor rigorously accurate. Nevertheless, it suffices for our present purposes and will help to answer a number of questions lurking in our minds.

At 100% relative humidity the air is saturated and cannot absorb any more water vapor. We know that water left in a dish will gradually disappear and that wet clothing will eventually dry. We also know that the rate at which this happens has something to do with the weather. That is because the air is unsaturated and absorbs water vapor. To hang clothes out to dry on a cold, foggy morning is a waste of time because the air is nearly saturated already. By contrast, the clothes will dry quickly on a hot, windy day.

If we could see the molecules at a water surface, we would see them jumping about. Some of them return to the water surface; others break free and remain in the air as vapor. As the air moves over the surface, it carries off a stream of molecules, which is why water evaporates more quickly in a wind. All this takes place without the water being anywhere near its boiling point.

Before the days of humidifiers, people commonly placed a dish of water over a register or on a windowsill. In the conditions of extreme winter cold prevailing over much of North America, the air drawn into a house from the outside holds very little water vapor because of its low temperature. On being heated from, for example, $-30°C/-20°F$ to $+20°C/+70°F$, its capacity for holding water vapor increases, yet its actual water vapor content remains the same as when it was drawn into the house at the outside air temperature. The relative humidity of the air thus becomes very low, and it becomes thirsty for water vapor, which it will draw by evaporation from any available source, such as wooden furniture and fixtures and the human body. This condition spoils furniture and makes life uncomfortable for inhabitants of the house, hence the dish of water or the humidifier.

When a jet of steam escapes from a kettle, the temperature of the steam emerging from the spout suddenly drops, with the result that water droplets condense as a cloud of visible steam. They mix, however, with the air surrounding the spout of the kettle, which is unsaturated. Because that air can accept additional water vapor, the droplets evaporate once again and the cloud of steam can be seen only around the spout of the kettle. Those of us who remember seeing steam locomotives at work can call to mind the difference between a hot, dry summer day when the white exhaust plume was almost nonexistent, and a damp winter day when the exhaust plume trailed far behind the locomotive. This was nothing more than a measure of the atmosphere's ability to reabsorb the water vapor condensed from the exhaust.

Clouds form because the air at that time and place is saturated with water vapor. The air reaches this saturated condition either through cooling or through the addition of water, or some combination of the two. Cooling occurs when air is lifted and expands, or when air passes over a cooler surface, or when the earth's surface cools at night by radiating heat into space and draws heat directly from the air above it. Water can be added when the air passes over an area of water, or when rain falls through it from above. If the air does not reach saturation, clouds will not form. Water vapor can condense either into water or directly into ice. The reverse is also true. Clouds, therefore, can be formed of water droplets, ice particles, or both. "Supercooled" water droplets are known to exist at temperatures well below freezing.

Important to the condensation process is the availability of tiny specks of material for water vapor to condense onto. If these specks, known generically as *condensation nuclei,* were not readily available, the air would reach a high level of supersaturation before condensation could occur. When condensation did occur, it would do so suddenly, resulting in the quick formation of clouds and cloudbursts. Fortunately, this is a matter for speculation only. Supersaturations of 400%–700% have been achieved only experimentally under carefully controlled laboratory conditions.

The specks of material in the atmosphere are surprisingly diverse and numerous. They consist of sea salt, volcanic dust, clay dust, soot from fires of all kinds, fly ash, bacteria, pollen, natural fibers, smokes, and other materials light enough to remain suspended in the air. They range in size from 0.002 microns up to 50 microns, and in concentrations from less than 300 particles per cubic centimeter to more than 100,000. (A micron is about 0.00004 inch; a cubic centimeter is about 0.06 cubic inches.) These specks act as nuclei onto which water vapor condenses from saturated air to form clouds.

If all this solid material exists in the air, why does it not fall? For that matter, if clouds are made up of droplets of liquid water, why do they not fall?

## Terminal Velocity

Cloud droplets and dust do fall. The dust that settles on everything in the still air of an unused room proves this. But the free atmosphere is continually churning and swirling and keeps these particles aloft. Any object falling under the influence of gravity through any fluid accelerates at 32 feet per second per second to a "terminal velocity" at which the object's weight equals the opposing force of the fluid's viscous resistance. High school physics courses demonstrate this with a ball bearing falling through glycerine. The falling object maintains its terminal velocity until it hits a surface. A skydiver's terminal velocity is about 125 mph. On jumping out of the aircraft, he accelerates to his terminal velocity and then free-falls at that velocity until he opens his parachute. The parachute suddenly increases the skydiver's total air resistance, with the result that he decelerates abruptly to a new terminal velocity at which he lands.

Let us compare the terminal velocities of some of the things that fall through air.

| Object | Terminal Velocity (ft/min) |
|---|---|
| Cloud droplet | 0.59 |
| Drizzle | 200 |
| Light rain | 390 |
| Heavy rain | 1,475 |
| ¾″ hail | 3,940 |

If air is rising at the same velocity as a water droplet is falling, the droplet will not fall but will remain aloft until it re-evaporates, grows larger by coalescing with other droplets, moves out of the rising air, or until the air stops rising. Let us look at some of the lifting mechanisms in the atmosphere.

| Mechanism | Lifting Rate (ft/min) |
|---|---|
| Warm frontal lift | 0.5–5.0 |
| Cold frontal lift | up to 10 |
| Strong cumulonimbus cell updraft | 5,000 |
| Orographic lift | 1,000–2,000 or more |

By comparing terminal velocities and lifting rates, we can see why clouds remain aloft, why frontal lift alone will not prevent rain from falling to the ground, and why cumulonimbus cells are able to recycle hailstones. Hailstones have been known to be as large as 5 inches in diameter. It is easy to realize that updrafts capable of throwing 5-inch chunks of solid ice around must be extraordinarily violent.

## Clouds

Clouds are among the most beautiful of all sky phenomena, and among the most mysterious and frustrating to the visual pilot. They are classified with long Latin names, and meteorologists seem to delight in subdividing them into smaller and smaller species with longer and longer names. For the visual pilot this classification can be replaced by one of perfect simplicity. Clouds are of two kinds: high and low. The high clouds forecast the weather; the low ones get in our way. We still have to use the basic Latin names, however, because they are as much a part of the language of aviation as flaps and ailerons.

The high clouds have names beginning with "cirro-" or "alto-." (There is "cirrus" but no "altus.") They are cirrus (CI), cirrostratus (CS), cirrocumulus (CC), altostratus (AS), and altocumulus (AC).

The low clouds are of the stratus and cumulus families. The stratus family consists of stratus (ST), nimbostratus (NS), and fog (F). The cumulus family consists of cumulus (CU), towering cumulus (TCU), and cumulonimbus (CB). The addition of the prefix "fracto-" or the suffix "fractus" means "broken up."

### HIGH CLOUDS

One of the standard tests for VFR flying weather is a piece of blue card with a hole in it. The pilot holds the card up to the sky. If the sky as seen through the hole matches the card, he takes off; if it does not, he stays on the ground. That is a well-known joke, but it contains an underpinning of sound theory. If the upper sky can be seen and is blue, the weather is basically fine. Why? Because the air high aloft is descending, thus becoming warmer and drier, or, at any rate, it is not ascending. High clouds show that the air at great heights is being forced to rise, cooling, becoming saturated, and forming clouds. This signifies the existence of a trough, front, or depression, which may bring foul weather to us on the ground or at lower altitudes.

Clear upper air does not rule out the possibility of fog, stratus, or local cumulonimbus activity, and these can be troublesome. Even so, blue upper sky is a definite comment on the overall state of the weather, and a favorable one. Significant cumulonimbus activity produces clouds at all levels; widespread fog and stratus also block off the blue sky, so the hole-in-the-card rule still holds good.

Some excellent VFR flying weather can exist beneath a high overcast. We should, however, be keen to find out why the overcast is there and be on the lookout for lowering and thickening of the overcast, precipitation beginning, and lower clouds forming. Flying conditions were ideal one quiet autumn day under a 9,000-foot altostratus overcast. An hour later the overcast had lowered to 5,000 feet. An hour after that the local airport was closed even to IFR traffic because of falling snow.

Highest and thinnest of the common cloud types is cirrus. Small patches of cirrus are common in high-pressure areas and are without significance. If, on the other hand, the cirrus thickens persistently from one side of the sky to the other, and if the thicker cirrus invades the whole sky in the course of a few hours, it is a certain forecast of some sort of weather system approaching. Typically, a change in the weather can be expected over the next 6–12 hours. The more slowly the clouds thicken, the weaker the weather system and the less low clouds and precipitation it will produce, and vice versa.

Sometimes we see bands of cirrus extending in straight lines across the whole sky. These mark a jetstream. A relationship exists between jetstreams and weather systems lower down, but it is neither immediate nor direct.

Cirrostratus and altostratus form a uniform white to light gray veil across the whole sky. They merge into one another and forecast the same things. They are typical of an approaching warm front, or they may be seen around the outer margins of a depression. At first the sun or moon shines through them, but with a halo. As the veil thickens, the moon disappears and the sun fades to a ball of light. At the "ball of light" stage precipitation commonly begins, continuing until after frontal passage, which may be expected another 6–12 hours after the start of precipitation. Low clouds may form in the precipitation. Once warm frontal precipitation begins, deterioration in the weather, possibly to below safe VFR limits, is imminent.

Because of the relative weakness of the moon's light, less

cloud is needed to hide it than to hide the sun. A halo around the moon indicates thin cirrostratus, and foul weather is still quite far distant. If, on the other hand, we know that the moon is up and neither it nor the stars are visible, we should check thoroughly on the weather before embarking on any wide-ranging night VFR flights.

On occasion the whole sequence of high clouds suggesting a warm front will appear. The sky then clears without either low clouds or precipitation, the whole sequence lasting perhaps 5–8 hours. This could mark the passage of a weak upper front, trough, or occlusion. The same sequence of events, although lasting for days rather than hours, might indicate the passage of a distant depression.

We are familiar with the classical surface pressure chart of a frontal depression with a warm front and a cold front radiating from its core. There is, however, the nonfrontal side of the depression, which appears, to someone on the ground, only as an area of cloud and precipitation. Several hundred miles from the center of the depression, all that may be seen is the outspread shield of high clouds. Lower and thicker clouds, perhaps with continuous precipitation, will be experienced at places on the ground closer to the core of the depression.

Cirrocumulus is not as common a cloud type as the others. When it does occur, it may be caused by turbulence between layers of air aloft.

Altocumulus is common, especially as the leading cloud of a front or depression. It also occurs as isolated patches in areas of settled weather and, as such, is harmless and portends nothing of significance. Altocumulus may have either a quilted or a dotted appearance. The difference lies in the degree of instability aloft. The more prominent the dots, the more unstable the upper air.

When the dots are very pronounced, they are called altocumulus castellanus. They look like miniature towering cumulus. Especially on a warm summer morning, they are a definite warning of thunderstorms, which can materialize as if by magic. Altocumulus castellanus indicates a layer of unstable air at about 8,000–15,000 feet above sea level. Under the influence of the sun's heat, a layer of unstable air may grow from the ground, deepening as the day progresses. When this layer merges with the existing unstable air aloft, a single deep layer of instability is suddenly formed, a situation ideal for thunderstorm development. Initially the bases of the storms may be high, but lower clouds may form in precipitation.

In one instance, a line of altocumulus castellanus at midday marked a trough of low pressure. Pilots crossing the line at 8,000–10,000 feet reported moderate to severe turbulence. By 3 p.m. this situation had developed into a line of thunderstorms 100 miles long with tops to 41,000 feet, producing 1-inch hail.

In spite of its innocuous appearance, altocumulus castellanus should be watched carefully.

Another type of altocumulus is the lenticular mountain wave cloud, altocumulus standing lenticularis (ACSL). Despite its name, this cloud is stratiform in appearance. It is discussed in Chapter 3 under the subject of mountain waves.

One other type of high cloud is worth mentioning, the contrails produced by high-flying aircraft. They tend to form, if at all, between 20,000 feet and the tropopause. They are caused by the local saturation of air by water vapor emitted with the engine exhaust. Sometimes we can hear or see an aircraft at great height, yet it is not producing a contrail, or only a very short one. At other times the contrail persists for hours as a spreading banner across the sky. Why? The answer is interesting to us as a long-range forecast. If a contrail is short-lived or absent, it is a sign that the air at that height is sinking and becoming warmer and drier, as is typical of anticyclones. The addition of moisture from the aircraft's exhaust cannot saturate the air for very long, if at all. Fair weather is likely. If a contrail persists, it is a sign that air aloft is rising and cooling as in prefrontal or cyclonic conditions. It is already close to saturation, and the additional moisture from the exhaust completes the saturation and cannot be reabsorbed for some time. This may warn us of frontal weather approaching over the next 24 hours.

This may not be the whole answer, because at any one time and place there are layers of air where contrails will form and others where they will not. Nevertheless, if we look out at the evening sky just before dusk and see a contrail, it may give us some protection against surprises the next day.

One morning the sky was cloudless with the promise of a fine day, except for a single broad contrail. By midmorning the sky was covered by a cirrostratus overcast, which lowered and thickened to become altostratus. Warm frontal rain began in the afternoon, giving way to a rainy evening with thick low clouds.

## LOW CLOUDS

The low, or at any rate low-based, clouds are the cumulus and stratus families. Cumulus-type clouds are lumpy or bulbous; stratus-type clouds are smooth and wispy. Stratocumulus is intermediate between the two and includes anything not indisputably one or the other.

Cumulus fractus — small puffs of cumulus — is common in the morning. It may evaporate during the day. If the air is moist and unstable, cumulus fractus may become cumulus, towering cumulus, or cumulonimbus. Cumulus fractus will bother us very little. We can fly around it, above it, or below it. Individual patches of cloud will not be more than a few hundred feet across.

Cumulus is more solid, as is the turbulence that can be associated with it. Individual clouds may be ½ mile to 1 mile across with 1,000 to 5,000 feet of vertical development. All cumulus clouds go through a similar life cycle of formation, growth, and decay. In the case of small cumulus, this cycle may be as short as 20 minutes. The sky may appear to be dotted with clouds, but observation will show that individual clouds are continually forming and dissipating. The same is true of thundercells, except that the cycle is longer — perhaps 45–60 minutes — and the results more spectacular. The reason for this cyclic nature is that the cell is formed by a local source of suitably conditioned air. This supply is soon exhausted. In the case of a small cumulus caused by thermals rising from a warm land surface, the cloud both shades the ground and exhausts the local supply of heated air. In the case of a cumulonimbus cell, the air is locally cooled and stabilized by the precipitation and downdrafts that the cell itself produces.

For the lightplane pilot the roughest ride is beneath a cumulus deck. Weaving between cumulus clouds is fun, but pilots can easily lose track of their whereabouts. Someone else may have the same idea, leading to the possibility of sudden confrontations. A risk of collision with IFR aircraft also exists; hence the rule for

2,000 feet of horizontal separation from cloud. This is easier said than done, either because the distance is hard to judge or because there may locally be less than 4,000 feet between clouds.

One summer evening, climbing northbound out of Toronto in a DC-9, a tiny, toy-sized aircraft could be seen flitting through great canyons of cumulus and towering cumulus. The "tiny" aircraft was a Boeing 747. Scale and distance can be illusory.

The smoothest ride is to be found above cumulus clouds, assuming that we can overtop them. Our view of the ground will be much reduced, hiding many landmarks, and navigation may have to be by dead reckoning or by electronic means. Cumulus can quickly grow past lightplane altitudes. Wise visual pilots ensure that the cloud does not become broken or solid beneath them.

Towering cumulus are larger than cumulus, darker gray, and often with ragged edges. If they have risen above the freezing level, their tops may have a fibrous appearance. Some of them show signs of cumulonimbus development, such as rain, hail, and strong turbulence. The difference between towering cumulus and cumulonimbus is slight. It lies mainly in whether the cell has reached the tropopause, which is where the anvil head forms.

All cumulonimbus clouds are capable of endangering a small aircraft in flight, regardless of whether they have or have not developed into thunderstorms. Those that have become thunderstorms are just that much more dangerous. A thunderstorm contains everything imaginable that is inimical to flight: extreme turbulence, violent up- and downdrafts, low-level wind shear, loss of visibility in the maelstrom of clouds and rain, airframe icing, hail, lightning, and sometimes tornadoes. Aircraft of all sizes and types have been lost to thunderstorms, torn apart by turbulence, engines extinguished by rain and hail, blown up by lightning, thrown out of control, iced up, thrown to the ground. Aircraft and thunderstorms do not mix. At one time it was thought that large and powerful aircraft could successfully penetrate thunderstorms, but losses went on and sound piloting has turned more and more to avoidance. Almost all weather books describe how thunderstorms work, so we will omit an explanation here.

For decades pilots found themselves looking up at thunderstorm tops, even though their aircraft flew higher in the course of time. The present record seems to be held by a U-2 pilot who saw a thunderstorm over India that he reckoned to top out at 100,000 feet. Storms reaching 30,000 feet are common, and violent storms in the American Midwest are known to punch through the tropopause to heights of 50,000–60,000 feet.

Individual storm cells are generally less than 10 miles across, but a severe storm can contain clusters of cells and in total be ten times that size. Updrafts of 5,000 feet per minute and downdrafts of 15,000 feet per minute have been measured. The cold outflow from a storm can act as a local cold front, triggering instability and propagating new storm cells.

The new pilot clearly should avoid thunderstorms at all costs. General practice is to avoid all thunderstorms by 10 miles and the severe ones by 20 miles. This is sometimes more easily said than done. Isolated air-mass thunderstorms surrounded by clear air do occur out on the prairies; by day they can be seen and avoided. But storm cells often are surrounded by decks of low- and midlevel clouds that hide the cells themselves. The cirrus anvil may be the most recognizable part of the cell. Even this is sometimes so ragged and drawn-out that it is not immediately recognizable for what it is. Thunderstorms are sometimes hidden in haze. Darkening skies, areas of heavy rain, increasing turbulence, and, above all, lightning are warnings to the VFR pilot. Some pilots have flown between cells only to find themselves boxed in by fresh cells forming around them.

Storms are notoriously difficult to predict; the means available for reporting their existence are more effective than those for predicting them. Charts of lightning plots, available at some weather stations, are useful. Arrays of detectors take cross-bearings on lightning strokes, and the information is plotted by computer onto a map.

The two common notes running through thunderstorm-related accidents are that the pilot flew into a storm that was not known to exist, or that the pilot penetrated a storm that was known to exist but that was more violent than anyone suspected. These two themes encompass fatal accidents to aircraft of all sizes from VFR singles to radar-equipped multiengine jets operating under IFR.

New pilots should consider avoiding an active thunderstorm

area altogether, especially at night or in mountain country, and especially if a SIGMET (see Chapter 12) has been issued. At night the visual pilot's main defense — vision — is nearly eliminated. The absence of lightning has proved to be no safeguard. In high mountains, routes often are restricted to valleys where the pilot risks being trapped when thunderstorms are active.

Towering cumulus and cumulonimbus may occur randomly in air masses above large areas of flatland, or they may occur in lines along fronts or troughs. Central Europe experiences a feature known as "thundery low," a small, self-contained, nonfrontal depression characterized by thunderstorms. Air-mass thunderstorms sometimes form lines over open country. Even the apparently random ones form in preferred localities, especially over isolated ranges of hills such as the Black Hills of South Dakota. If a forecast calls for "risk of" or "occasional" towering cumulus or cumulonimbus, it means that these clouds will develop first in areas most favorable to their formation. This is a piece of local knowledge worth finding out. Abundant literature exists on how thunderstorms work, to which the reader is earnestly recommended.

Thunderstorms over land tend to follow a diurnal cycle, with maximum activity in the middle of the afternoon. Strongly developed afternoon thunderstorms may continue into the night. In some cases, however, water that is warmer than the land causes instability, which can touch off thunderstorms at any hour of day or night. Nocturnal thunderstorms off the coast of Florida are caused in this way. The thunderstorms seen on summer nights in the bush country of Northern Ontario, Wisconsin, and Minnesota, with its numerous lakes, may also stem from this cause. In the Pike's Peak area of Colorado, the air in summer, though dry, is so violently unstable that thunderstorms can occur almost incessantly at any hour of the day or night. Thunderstorm activity along a front is likely to vary through the 24 hours in response to the daily cycle of stability. (For more on stability, see Chapter 9.) The maps in Figures 7.1–7.4 give average numbers of thunderstorm days in the U.S. for each of the four seasons.

Stratus is a smooth, shapeless cloud type, white to gray in color depending on its thickness and the thickness of cloud cover above it. Precipitation, if any, will be drizzle. Whereas cumulus-

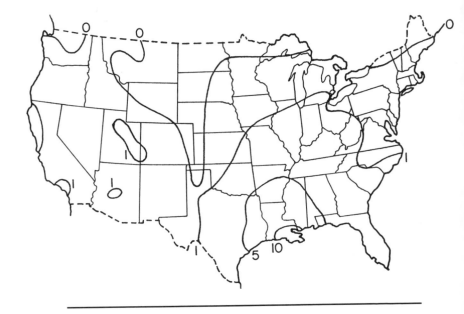

Fig. 7.1. Average number of thunderstorm days in season: Winter. Adapted from *U.S. Air Force Weather Manual* (AFM 51–12).

type clouds show a marked distinction between clouds and clear air, the same is not true of stratus-type clouds, which can range from thin mist to thick cloud. This adds to the risk of entering such clouds inadvertently in poor visibility. Stratus can be indistinguishable from other stratiform clouds, rain, calm water, or unbroken surfaces of snow and ice. Drizzle or snow, mist, and low stratus add up to a dangerous sky condition for the visual pilot. The problem is compounded many times by darkness, not least because stratus and fog often form at night.

Nimbostratus is a cloud type characteristic of warm fronts.

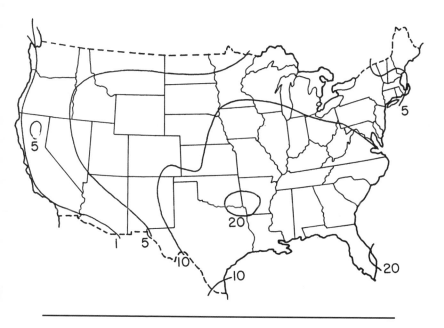

Fig. 7.2. Average number of thunderstorm days in season: Spring. Adapted from *U.S. Air Force Weather Manual* (AFM 51–12).

It fills the sky with a shapeless mass of clouds, which may be continuous from near the ground to near the tropopause and may stretch unbroken for 200 miles across the front and as much as 1,000 miles along the front. An area of nimbostratus may also occupy the core of a mature frontal depression without being associated directly with a front. Such clouds are caused by the widespread uplift of moist, stable air. Because of its great depth, nimbostratus is dark gray underneath and probably will produce continuous rain or snow.

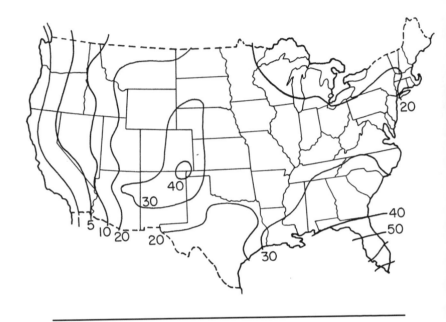

Fig. 7.3. Average number of thunderstorm days in season: Summer. Adapted from *U.S. Air Force Weather Manual* (AFM 51–12).

Fog is the absolute bane of aviation. It is stratus that has formed at ground level. The European definition of fog calls for surface visibility to be less than 1,000 meters (3,280 feet). In North America this definition is not used. We may usefully think of "thin" fog and "thick" fog.

Thin fog is typical of rainy weather under low overcast skies. An airport weather report might include: "Ceiling one thousand overcast, visibility four miles in light rain and fog." The observer reckons that he can see that far, but his vision is restricted by rain

Fig. 7.4. Average number of thunderstorm days in season: Fall. Adapted from *U.S. Air Force Weather Manual* (AFM 51–12).

and by a general obscuration, which he calls fog. The epitaph of all too many pilots, "Continued VFR into IFR weather conditions," is a product of fog such as this.

Thick fog tends to form in well-defined patches. Inside the fog bank, visibility may well be less than a mile and may be close to zero. Outside the fog bank, visibility may be unlimited with clear skies overhead. The map in Figure 7.5 shows annual days of thick fog in the U.S.

Three types of fog are most common in the inhabited parts of the world: radiation fog, advection fog, and upslope fog. Ice

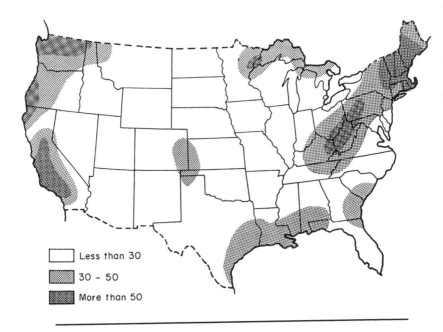

Fig. 7.5. Annual days of thick fog (visibility less than 400 meters). Source: M. Neiburger, J. G. Edinger, and W. D. Bonner, *Understanding Our Atmospheric Environment* (San Francisco: W. H. Freeman, 1982).

fog and "Arctic smoke" are specific to high latitudes.

1. *Radiation fog.* This type forms at night in calm weather under clear skies. The earth radiates heat into space and draws heat from the air directly above it. If the air is humid enough, this cooling results in condensation and, hence, fog. If the air is saturated by rain during the day and the sky then clears in the evening, and if winds are calm, it is an ideal situation for radiation fog to form. Air cooled by contact with the earth's surface will drain into valleys or areas of low ground, and this is where fog will form in the early morning.

Fog often becomes deeper and thicker during the first few hours after sunrise. This is attributed to slight thermal churning of the air by the first heat of the sun. The fog then "burns off" later in the day. In summer, the burning off may be complete quite early in the day. In autumn, as daybreak comes later and the sun's heat as received at the earth's surface weakens, the fog burns off progressively later in the day. In winter, cities such as London, Paris, Vancouver, and Seattle are sometimes fogged in for days on end. If a layer of high cloud overlies the area, the burning off may be completely prevented. Wind either prevents radiation fog from forming or disperses it once it has formed.

2. *Advection fog.* When moist air flows over a colder surface and is cooled to its dewpoint, advection fog forms. The important distinction, therefore, between radiation and advection fog is that radiation fog does not form when a wind is blowing, whereas advection fog is produced and accompanied by wind. The parts of North America best known for advection fog are the coasts of California and Newfoundland. In both cases the fog is caused by contrasts in seawater temperatures just off the coast. In California onshore winds bring warm, moist air from the Pacific over cold, upwelling water along the coast. In Newfoundland air warmed by the Gulf Stream produces fog when moved over the cold, southward-flowing Labrador Current.

Whereas radiation fog, at least in general terms, follows a daily cycle that can be predicted to some extent according to the time of year, advection fog is caused by the regional wind circulation and follows no daily pattern. On the south coast of Newfoundland, for example, it is well known that a southeast wind brings fog. The fog, being produced continuously in air streaming over the Labrador Current, moves over the land, penetrates for several miles inland and then gradually disperses by evaporation over the land. As long as the pressure distribution causing this flow remains in place, the fog persists, sometimes for a week at a time. As soon as the wind shifts, the fog stops forming. Banks of either type of fog can lift to form a low stratus overcast.

3. *Upslope or hill fog.* This type forms when moist, stable air moves over rising ground and is cooled to its dewpoint by adiabatic expansion. Upslope fog is common in mountainous areas. It is, however, most widespread over large areas of gently

rising ground, such as that between the Gulf Coast of Texas and the plains of eastern Colorado. This type of fog, too, is a function of wind circulation rather than of diurnal cycles.

Fog, whether thin or thick, is not for the visual pilot. Thin fog is an almost certain warning of IFR conditions. If airport observers are reporting a visibility of 3–5 miles in fog, that is as far as they can see. Ten miles away ceiling and visibility could be close to zero and the pilot would not know it. "Marginal VFR" in these conditions is to be avoided. In thick fog we cannot even see to take off, but, approaching a fogbank in flight, we have several options.

Fog often is less than 5,000 feet deep unless it is associated with other cloud types. Therefore, VFR flight above a shallow fogbank is entirely feasible. The ground should be visible within gliding distance, which means that we can safely cross a fog-filled valley or strait 10 miles wide at 5,000 feet above the fog. Flight over extensive fogbanks, whether legal or not, is unwise in case we should need to land in a hurry.

Trying to fly beneath a fogbank is suicidally dangerous. It exposes the pilot to every possible danger of hitting obstructions, losing his way, or becoming disoriented. The fog often moves about slightly, lifting and settling, and the pilot, once under a fogbank, cannot be certain of an escape way remaining open behind him.

The only safe way to bypass fog is to fly around it or to climb as high as we can and try to see the ground on the far side of it. If we cannot do either of these things, we have to turn back. Any problems caused to the visual pilot by fog are greatly increased by darkness.

One type of stratiform cloud seen in mountain country which does not fit these categories is the eddy cloud. It is seen as a bed of ragged stratus filling the lee side of a ridge. Wind blowing over the ridgecrest sucks air up the lee side, which, in suitable conditions, condenses to form the eddy cloud.

## Haze and Smoke

Although haze and smoke are not cloud types as such, they can severely restrict visibility. They are especially common and thick in the Northeastern United States and in the San Francisco and Los Angeles areas. Where forest industries are active, slash burning produces abundant smoke. Slash, the scrap material from logging, usually is burned in a period of calm weather following the first rains of autumn. The lower atmosphere can become a turgid mass of haze, smoke, and leftover fog patches. There is a risk of loss of horizon over large areas of calm water.

Forest fires should be avoided, as they pose a risk of disorientation in smoke and collision with firefighting aircraft. Large fires generate severe turbulence. Forest fires are reported in NOTAM (Notice(s) to Airmen).

Haze and smoke make navigation difficult and other aircraft hard to spot. The VFR pilot staring at the ground and the IFR pilot staring at his instruments in conditions intermediate between IFR and VFR are not devoting much time to looking for traffic. Landing lights and strobes help to make the aircraft more noticeable. But dense traffic in haze is something to avoid, if possible.

## Weather Reports

Aviation weather reports describe clouds in terms of the altitude of the base of each layer and the amount of the sky they cover. In *surface actuals* and *terminal forecasts,* the heights are in feet above station ground level. In *area forecasts,* because of the variations in terrain heights within the forecast area, cloud heights are given in feet above sea level unless otherwise stated. The area forecasts predict cloud-top heights while the other two do not.

The degree of cloud cover is given as "clear," "scattered," "broken," or "overcast." "Clear" and "overcast" are easy to visualize. "Clear" is what we like to hear best of all. "Overcast" means that the sky is completely covered by clouds at whatever height is given. "Five thousand overcast, visibility fifteen miles" means a defined high ceiling with good visibility beneath and, as such, good VFR flying conditions, provided that wind speeds are reasonable. "Scattered" means a cloud layer covering anything up to half the sky. We can climb above a "scattered" layer or weave between clouds. Where clouds are scattered, they often are associated with terrain features such as hills, islands, or shorelines which have in some way caused the cloud to form. "Broken" means a cloud layer covering more than half the sky but not completely overcast. The breaks in a broken layer can be few and far between. The base of a broken layer is defined as a ceiling, and consequently our operating VFR ceiling is 500 feet below it.

One feature of the sky condition reports that is not obvious is the principle of *summation opacity.* In a surface actual report the weather observers add the tenths of sky coverage of each layer. A layer of two-tenths at 500 feet and another two-tenths at 1,500 feet add up to four-tenths and would be reported as "500 scattered, 1,500 scattered." Three-tenths at 500 feet and three-tenths at 1,500 feet would be reported as "500 scattered, ceiling 1,500 broken" because the two layers add up to a "broken" sky condition in total even though the two layers separately are "scattered." Another two-tenths at 2,000 feet would be reported additionally as "2,000 broken" because the total is still only eight-tenths and the observer can see blue sky. If there were four-tenths at 2,000 feet, the whole assemblage would be reported as "500 scattered, ceiling 1,500 broken, 2,000 overcast."

VFR pilots flying above a scattered layer must be sure that it does not become broken or overcast beneath them. If it did, they could legally continue "VFR on top" in the United States but not in Canada. To argue the why and the wherefore of these regulations is beyond the scope of this book, but the visual pilot flying VFR on top should be aware of certain things.

First, the cloud deck beneath must open up in the area of the destination so that the plane can descend through it without entering clouds. If pilots choose to fly over a layer with few breaks in it, they must accept the fact that any emergency descent would have to be made through clouds. If such a descent were the result of loss of power, the attitude and direction gyroscopes, which are commonly powered by an engine-driven vacuum pump, would be running down and these two instruments would be giving false readings. It is important that VFR conditions and sufficient unobstructed airspace do, in fact, exist beneath the clouds near the destination.

Second, the tops of cloud decks may rise beneath the aircraft "on top," forcing the pilot to climb either until the aircraft's ceiling is reached or until oxygen is needed. This may be because the cloud tops are rising or because the aircraft is flying over progressively higher clouds. If these conditions become apparent, pilots have no safe option but to turn around, hoping that cloud tops are lower behind them.

Third, if pilots fly between layers, the layers may merge, or they may encounter other layers between the first two, or they may enter instrument conditions because of precipitation falling from the upper layer. At night, stars or the moon may provide enough light to use the top of a cloud deck as a horizon reference if there are no higher clouds. Visual flight between layers would become instrument flight with the onset of night. "VFR on top" means just that, flight above all clouds. VFR on top can be a useful way of doing things, but it has snags and must be used with care. The situation is not always clearly definable.

Visual pilots are in no way obliged to cower on the ground just because there are a few clouds about. On the other hand, it is their business to know why the clouds are there and what weather they portend.

# 8

# Precipitation
## "When the Great Goose Moults" and similar legends

**PRECIPITATION INCLUDES** any form of falling water in the sky. Best known are rain, snow, drizzle, and hail. Precipitation also occurs in forms such as ice pellets, which most people do not distinguish from snow or hail. A strong correlation exists between extensive continuous precipitation and bad VFR flying conditions. In fact, it should be apparent already that high winds, low clouds, and precipitation all add up to bad flying weather. They very often accompany one another, and one may forecast the others. Precipitation can range from the harmless to the dangerous, so let us have a look at the different kinds.

The earlier part of this book does not discuss precipitation because it seldom has direct effects on lightplane performance. Even the most intense rain associated with thunderstorms and hurricanes has only rarely affected the performance of piston-engine subsonic aircraft. The visual pilot, however, has many reasons for avoiding intense rain. The most significant effect of precipitation for the visual pilot is to impair visibility. It also serves as a warning of changing weather conditions, which may have a similar effect. The exception to this statement is hail, which can conveniently be mentioned here.

## Hail

Hail consists of small (and sometimes not so small) balls of solid ice, commonly the size of peas or ball bearings. It is formed by water droplets being lifted by updrafts above the freezing level, falling out of the updraft, being lifted again, and collecting ice as they move about inside the convective cell. The necessary combination of water and updrafts is found in towering cumulus and cumulonimbus cells. The stronger the updrafts, the more times the hailstones can be recycled in the cell, and the larger they can become. Eventually the hailstones are ejected from the convective cell and fall outside it. Or they grow to such a size that the updrafts can no longer lift them, or the updrafts weaken, allowing the hailstones to fall to the surface. Hail most often falls ahead of the advancing cell, commonly under the cirrus anvil. It may be visible as a white veil or streamer, frequently from considerable distances if lit by the sun.

Flight into hail makes a noise the pilot is unlikely to forget, and an immediate 180° turn is the recommended action. Hail will at the very least take the paint off leading surfaces. It may leave dents to the extent of seriously deforming the leading edges of wings. Windshields, light glasses, and radomes are especially vulnerable to hail damage.

## Rain

So why does it rain? Many people have spent much time flying about inside clouds doing the scientific equivalent of holding a bucket out of the window to find the answer to this question.

Raindrops are believed to form in two ways: by coalescence and by freezing. Cloud droplets are of various sizes with various terminal velocities. They therefore collide and coalesce, resulting in drops large enough to fall out of the cloud. The stronger the lifting of the air, the faster is the rate of condensation and the heavier a raindrop must be to fall through the rising air. That is why strong uplift along active fronts, above mountain ranges, or in convective cells produces more and heavier rain than does the weaker lifting that characterizes quieter conditions.

The freezing process occurs when ice crystals form above the freezing level. Water vapor condenses onto them and they start to fall as snowflakes. Below the freezing level they melt and reach the surface as rain. Which of these processes is dominant depends on the air temperature inside the cloud and, hence, on the latitude and season of the year.

In temperate latitudes much of the precipitation reaching the earth's surface as rain is formed as snow. If the lower air is cold enough, it does not melt but, rather, falls to the surface as snow. Sometimes precipitation does not reach the surface at all; instead it evaporates after falling out of the clouds.

Light rain will not, of itself, seriously impair visibility. Moderate rain can reduce visibility to 3 miles or so. Intense rain may reduce visibility to a mile or less. Saturation of the lower air often causes thin fog or patches of stratus to form that are difficult to see in the generally impaired visibility. Thick cloud cover must be present to generate continuous moderate or heavy rain. Such conditions accelerate the onset of darkness, accompanied by increasing low clouds. Rain showers can be flown through or avoided, but widespread continuous rain can make navigation difficult. Low, ragged cloud bases, rising terrain, approaching dusk, and combinations thereof, can alter this situation from the merely inconvenient to the hazardous.

Rain is a useful warning of other weather conditions. If rain begins, especially if reasonable weather was forecast, the visual pilot should be wondering: "Why is it raining? How does this tie in with the forecast? Is the forecast correct? How much worse is the weather likely to become?"

A typical example might concern the arrival of a front. Suppose that a weather briefing ran, in part: "There is a front coming in, but it is not due to reach here until tonight. It is reported to be fairly weak anyway." Now let us suppose that it is midmorning and that, as we preflight the aircraft, we notice rain beginning to fall from a high overcast. Soon the ground is wet under a steady, light rain. The rain itself is no problem, but the immediate implication is that the forecast is wrong. The effects of the front are beginning to arrive now. It is moving faster and is more active than anyone had realized. The weather certainly will become worse, and within the next 4 or 5 hours conditions may well deteriorate past what we can accept.

## Drizzle

Drizzle is a form of rain in very small droplets. It is the characteristic precipitation of stratus-type clouds. Described accurately as "misty rain" in some parts of the world, it can obscure visibility. Stratus, drizzle, and thin fog often occur together and blend into one another. The combination does not constitute VFR flying weather. Add mountains, dusk, or an extensive area of water or snow, and we have a situation in which loss of outside horizon reference is likely. Such a situation is extremely dangerous to the new pilot.

## Freezing Rain and Ice Pellets

Frontal rain is produced by the lifting of warm air over colder air. If the temperature of the cold air is below freezing, and if the warm air is warm enough for precipitation to be falling in it as rain, the rain does not, on entering the cold air, become snow as we might suppose. It remains for a time as supercooled water. As such, it will freeze on impact with any solid object and coat it with ice. This is called freezing rain and is extremely dangerous to aircraft in flight because the rate of ice accretion exceeds that of any other icing condition. If the falling rain remains in the cold air long enough to freeze before reaching the ground, the result resembles soft hail and is called ice pellets.

## Snow

Snow reaches the surface when the whole precipitation process takes place at temperatures below freezing. Wet snow can stick to an aircraft in flight with a tendency to choke air intakes. The most serious effect of snow, however, is that it drastically curtails visibility. Snow can begin to fall over wide areas simultaneously. Heavy snow showers can reduce visibility from several miles to ¼ mile or less in 5–10 minutes, closing airports instantly, even to IFR traffic.

If there is already snow on the ground, loss of horizon reference is a potentially severe hazard if snow begins to fall. If visual pilots encounter falling snow, they should head for the nearest airport, being careful to avoid flying over areas of unbroken snow or large expanses of frozen water. In poor light a uniform terrain, such as forest, does not provide a good horizon reference. The primary requirement in falling snow is to avoid becoming lost and to keep adequate horizon references in sight. When "snow flurries" are forecast, the continuity of cloud cover will indicate how widespread any snowfall is likely to be. Snow flurries from isolated cumulus clouds can be avoided easily enough. Flurries from a broken or overcast stratocumulus deck, however, could be more widespread and harder to avoid. In spring, melting snow beneath an overcast, rainy sky is likely to form thin fog. Low ceilings, poor visibility, and snow-covered ground are a dangerous combination for the visual pilot.

## Sequences of Precipitation

Sometimes the sequence of precipitation types is enough to tell us what is going on with no other information. Suppose — and this is no improbable supposition — that the only information we have from looking at the sky is that we see a continuous and persistent gray overcast. As night falls we cannot even see that. During the afternoon light snow falls. About dusk this turns to ice pellets. Later in the night we notice that the air is markedly warmer and a steady rain is falling. This is a typical warm frontal sequence.

In the afternoon we are in the cold air mass. Precipitation originates above the frontal surface in the "warm" air as snow and remains as such in the cold air. As the front approaches and the frontal surface lowers overhead, precipitation from the warm air melts to form rain but then freezes into ice pellets on falling through the cold air. After the front passes, we are in the warm air mass and precipitation reaches the ground as rain. In the same way, continuous light rain or drizzle, followed by showery precipitation and an increase in wind speed, would be typical of a cold frontal passage otherwise concealed in a continuous overcast.

These are weather conditions as they actually occur, rather than the classical cloud sequences of the meteorology textbooks. If new pilots miss no opportunity to observe the weather from the ground, they will be subject to fewer surprises in the air.

## Whiteout

Whiteout is a condition that can develop between a snow-covered land or ice surface and an overcast, with or without snow falling. To amplify our previous remarks on this subject, Arctic travelers have found that "these whiteouts obliterated all visual clues of depth, height, and distance and made us feel as though we were suspended in skim milk. The maxim of polar travel when whiteouts develop is to make camp and sleep until visibility returns" (*National Geographic,* 1986, 308). If that is true when traveling on the earth's solid surface, whiteout obviously presents an impossible situation for any pilot, instrument-rated or not. Even an instrument pilot needs clear visual reference in the last 15 seconds before touchdown. Wherever and whenever whiteout is forecast, reported, or suspected, the visual pilot's place is on the ground.

Precipitation tends not to affect the aircraft's performance directly. It is, however, a valuable indicator of sky conditions and a warning of conditions to come. Precipitation, especially snow, does obscure visibility and exposes the new pilot to the risk of becoming lost. Precipitation often combines with, or causes, other factors to obscure visibility and produce a hazardous situation for visual flight.

# 9 Atmospheric Stability
## What many talk about but few understand

**THE STABILITY** (or instability) of the air overhead is one of its more important and pervasive characteristics. It affects all sorts of phenomena such as clouds, turbulence, precipitation, visibility, and the behavior of weather systems and air masses. Atmospheric stability is a matter of lapse rates. Unfortunately, confusion abounds on the subject of lapse rates and, hence, on that of stability.

## Lapse Rates

The part of the atmosphere in which we live and fly is called the *troposphere*. In the troposphere the air grows cooler with increasing height (with certain exceptions). The rate at which it grows cooler, in degrees of temperature per thousand feet of height, is called the *lapse rate*. Above the troposphere is a zone called the *stratosphere,* in which the temperature is almost constant with increasing height. The boundary between the troposphere and the stratosphere is called the *tropopause*. The height of the tropopause varies from a low of 25,000 feet over the Arctic in winter to 50,000 feet over the equator.

The air temperature at the tropopause over the equator is much colder, at −75°C/−103°F, than over the poles, at −45°C/ −49°F. Over the middle latitudes the tropopause is at a height of 30,000–35,000 feet and the air temperature there is −56°C/ −69°F.

Some aircraft instruments are affected by the temperature of the air. To allow these instruments to be calibrated, a reference lapse rate has been calculated to represent the troposphere at about 45°N latitude. Called the *ICAO Standard Lapse Rate,* this is taken to be 1.8°C per thousand feet.

Air temperatures at various heights are measured by means of a balloon carrying instruments. This device is known as a radiosonde. The resulting graph of temperature versus height is called a *sounding curve*. The sounding curve at some time and place may show a constant decrease of temperature with increasing height, which equals the ICAO Standard Lapse Rate. More often it will not. In all probability, the sounding curve will show layers of air with steeper or flatter lapse rates or, indeed, layers that become warmer with increasing height, called *inversions*.

## Thermodynamics

If air is compressed, it becomes warmer; if it expands, it becomes cooler. Two laws of thermodynamics — Boyle's and Charles's — combine to state that, for a gas, pressure times volume divided by temperature is a constant. Thus, if one is changed the other two will compensate for it so the net result is the same.

Common expressions of this law can be found in the operation of compressed-air equipment. When a compressor is running, the high-pressure air outlet becomes hot. The air drawn in from the atmosphere is being compressed to a smaller volume and higher pressure, becoming hotter as a result. In fact, in a construction-type compressor the limitation on continuous output pressure is not the ability of the machine to withstand the pressure but, rather, its ability to dissipate the heat generated by compressing the air. At the other end of the compressed air line the compressed air driving the tool, such as a rockdrill, is suddenly released to atmospheric pressure upon escaping through the tool exhaust, and it cools suddenly — to such an extent that moisture in the air sometimes condenses out and freezes around the exhaust port. In the same way, when the compressed carbon dioxide in a fire extinguisher is released to atmosphere, it cools instantly and makes the nozzle so cold that frost forms on it.

The graph plot of temperature versus pressure in these cases is called a *process curve.* If heat is not exchanged between the gas and its surroundings, the process is said to be *adiabatic.*

## Adiabatic Cooling and the Dry Adiabatic Lapse Rate

If air is forced to rise, as when blowing over a range of hills, its pressure is reduced with increasing height, and the air becomes cooler as a result. This process is called *adiabatic cooling* and the graph plot of temperature versus pressure would be a process curve as defined above. Meteorologists have expressed this process in terms of degrees of cooling per thousand feet of lifting and have called it an *adiabatic lapse rate* because it is convenient to do so. The reason why it is convenient is so that the process curve of the adiabatic cooling caused by lifting can be compared directly with the sounding curve or lapse rate of the air before it is lifted. This shows how stable the atmosphere is at some time and place — an important aid to weather forecasting.

Let us imagine air blowing over open plains on which stands an isolated hill. The moving air has its own sounding curve, which can be measured by means of a radiosonde or from an aircraft. Where the air strikes the hill it is forced to rise; in doing so it expands and cools adiabatically. This cooling process is independent of the sounding curve in the undisturbed air flowing past the hill on either side. Thus, at the elevation of the top of the hill two lots of air are side by side: the undisturbed air and the lifted air. The undisturbed air at that elevation is at one temperature according to its sounding curve. The lifted air is at another temperature resulting from the adiabatic cooling process. The *dry adiabatic lapse rate* is 3°C per thousand feet. Thus, if the air started off at 15°C and was lifted 2,000 feet by flowing over the hill, its temperature will be reduced by that process to 9°C. If the sounding curve in the unlifted air happened to equal the ICAO Standard Lapse Rate, then the unlifted air level with the hillcrest would be at 11.4°C. More likely it will not.

At this elevation the lifted air may be warmer or cooler than the unlifted air at the same elevation. If the lifted air is cooler, and hence denser, it will flow down the lee side of the hill and revert to its former temperature. The lapse rate of the sounding curve is less than the dry adiabatic lapse rate and the air is said to be *stable*. If the lifted air is warmer than the unlifted air at the same elevation, then, being less dense, it will go on rising of its own volition and will go through further adiabatic cooling. In this case the lapse rate of the local sounding curve is greater than the dry adiabatic lapse rate and the air is said to be *unstable*. The stability of the air can be determined by comparing its sounding curve with the adiabatic lapse rate, of which the "dry" rate is but one. This information helps us predict the air's behavior.

## Saturated Adiabatic Lapse Rate

Most air, as we know, contains water vapor. If the air is cooled enough it becomes saturated and the water vapor condenses out as droplets, which form clouds and may form precipitation. When a substance changes state from solid to liquid or liquid to gas, extra energy is needed to accomplish the change of state. When the substance changes back from gas to liquid or from liquid to solid, the energy is released again. This increment of energy is called *latent heat*. If we spill gasoline on our hands it evaporates and feels cold because it is drawing the necessary latent heat from our skin.

When air is forced to rise to a height such that adiabatic cooling causes condensation, the air is warmed by the release of latent heat and its rate of cooling with increasing height is reduced. This new rate is called the *saturated adiabatic lapse rate*. The altitude at which condensation takes place is called the *lifting condensation level.* If air is forced to rise farther, above the lifting condensation level, it will cool at the saturated adiabatic lapse rate, which is only about 1½°C per thousand feet.

In the case above, let us suppose that in being lifted from sea level to 2,000 feet, the lifted air cooled adiabatically at 3°C per thousand feet to 9°C and that condensation took place. A cloud base would form on the hillside at 2,000 feet. If the hill were 4,000 feet high, the air would be lifted another 2,000 feet at the saturated adiabatic lapse rate and its temperature would be reduced by another 3°C to 6°C. If, by being lifted to that height, the air were to become warmer than the neighboring unlifted air, it would go on rising and 4,000 feet would be the *level of free convection.* The air would be said to be conditionally unstable. In other words its instability would be conditional on condensation and further lifting taking place. Figures 9.1 and 9.2 show various conditions of stability.

## Sources of Lifting

The altitude of the lifting condensation level and the elevation of the top of the hill determine whether the hilltop is covered by clouds or whether a cap cloud floats above the hill. At normal atmospheric pressures air behaves as if it were incompressible. As a result, the air forced upward over the hill forces the air above it to rise to greater heights above the hill itself. A cap cloud therefore may form in air which does not actually strike the hill. If the air is dry, no cloud will form.

A hill or a range of mountains is only one cause of air lifting and is called orographic lifting. Another is frontal lifting which occurs as warm air is wedged up over colder air. Convergence is another; this occurs when airstreams meet at surface. A small-scale example is the meeting of sea breezes over a peninsula or island. Convergence on a larger scale occurs in depressions as air spirals inward toward the core of the depression and upward to diverge at high altitudes. Troughs and depressions are thus features with the potential for showery, thundery weather. Such large-scale convergence can cause masses of clouds which are continuous for hundreds of miles and extending to heights of

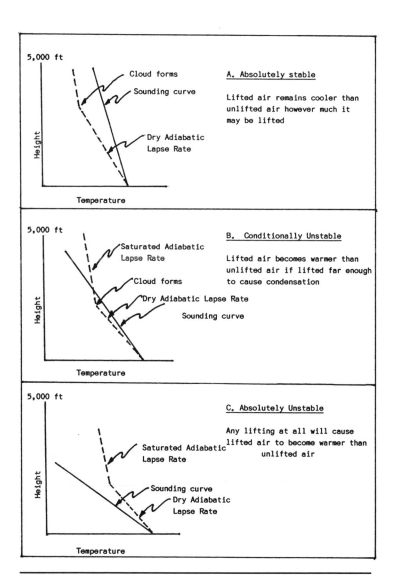

Fig. 9.1. Atmospheric stability: lapse rates and sounding curves (5,000-foot altitude shown for scale only).

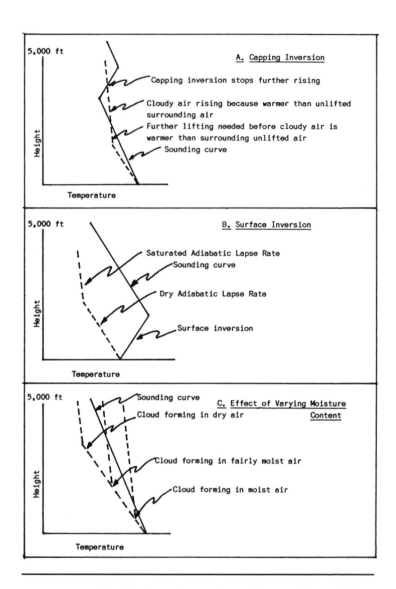

**Fig. 9.2.** Examples of various stability conditions (5,000-foot altitude shown for scale only).

20,000 or 30,000 feet. Conversely, an anticyclone is an area of air gently subsiding from high altitudes to diverge at surface. This suppresses lifting and accounts for the stable conditions characteristic of anticyclones. Air can also be lifted by local heating of the ground, which makes one patch of air less dense than its surroundings. As the bubble of warm air rises, adiabatic cooling takes place and cumulus clouds are a typical result. Turbulence is also a lifting mechanism. The airflow may be disrupted by ground features or by one layer of air shearing over another. Extensive stratocumulus decks are a common product of this lifting mechanism.

The moisture in the air forms cloud types which are characteristic of stable or unstable conditions. Stable air forms stratus-type clouds; unstable air forms cumulus-type clouds. The amount of moisture in the air and the height to which lifting or free convection go on determine the amount and type of clouds.

## Limits of Free Convection

Two intriguing questions arise: 1. Once launched in unstable air, why does a bubble of rising air not go on rising forever and what stops it? 2. What makes air stable or unstable at some particular time or place?

Several things can put a lid on free convection. Bearing in mind that the sounding curve is only rarely a straight line, there may be stable layers aloft or even inversions. Such a layer will bring free convection to a halt. Sometimes this can be seen when the tops of cumulus clouds (forming in an unstable layer) merge with higher stratus (forming in a stable layer). The strongest stable layer of all is the stratosphere. Even so, violent thunderstorms can penetrate the stratosphere for several thousand feet before their upward progress is brought to a halt.

Water droplets in the top of a growing cumulus cloud may radiate heat faster than they absorb it from the surrounding air, and faster than heat can be supplied by condensation. This can restrict the cloud's further growth.

The top of a building cumulus cloud may reach a layer of dry air into which its crest evaporates. This causes the rising air to revert to the dry adiabatic lapse rate, thus increasing the likelihood that it will come into equilibrium with its surroundings.

An important fact not usually mentioned in basic meteorology courses is that the saturated adiabatic lapse rate is not constant. We saw in Chapter 7 that the amount of water vapor the air can hold depends on its temperature. Moreover, the relationship between the saturation mixing ratio and temperature is almost an exponential one. The amount of latent heat released on condensation depends on the mass of water vapor being condensed. If the air is cold it cannot hold much water vapor and the warming effect that causes the saturated adiabatic lapse rate to differ from the dry adiabatic lapse rate is thereby reduced. The result is that the saturated adiabatic lapse rate gradually increases with decreasing temperature, so that at temperatures below $-29°C/$ $-20°F$, the dry and saturated rates are almost the same. In the ICAO Standard Atmosphere this temperature would be reached at an altitude of about 25,000 feet. As we might guess, that is not the whole story, but it will do for now. To learn more, the reader is referred to Chapters 7 and 8 of Horace R. Byers's *General Meteorology.*

The gradual decrease of the warming effect of condensing water vapor with decreasing temperature tends to limit the heights to which convection in unstable conditions can go on. Thunderstorms that grow beyond 30,000 feet are the product of hot air laden with moisture and intense instability extending to great heights, which brings us to our next question.

## Causes of Stability and Instability

Why should air be stable or unstable? We know that air is heated and cooled from below by contact with the earth's surface and that the temperature of the surface varies from time to time and place to place. If the ground cools or the air moves over an area of cooler land or water, its bottom layer is cooled. The more it is cooled, the more stable it becomes. If the ground warms or the air moves over an area of warmer land or water, its bottom layer is warmed. The more it is warmed, the more unstable it becomes. The sharper the decrease of temperature with height, the more likely it is that lifted air cooled by the adiabatic process will be warmer than the unlifted air surrounding it. The reverse is also true. Thus "stability" is a measure of how much lifting is needed before free convection will occur. Meteorologists determine this by plotting radiosonde measurements on a gridded sheet of paper called a *tephigram.*

Other things being equal, there is a cyclic daily change in stability. During the night the earth's surface becomes cooler. The bottom of the atmosphere is cooled by losing heat to the earth's surface and becomes more stable. On a cold, still night an inversion may form near ground level. As solar heating builds up during the day, this stabilizing process is reversed and the air becomes less stable. Mixing by wind and convection will spread this instability upward. Sometimes a stratus layer can be seen to change to stratocumulus during the day, reverting to stratus in the evening. This is an example of small-scale destabilization during the day.

The temperature of the land or water over which moving air passes affects its stability. Cold air passing over a warm surface is warmed from below and destabilized. Thus, cold air moving over the Great Lakes in winter not only picks up moisture but becomes unstable because of the relatively warm lake water. Especially when combined with frontal lifting, this can produce heavy snowfall and even thunderstorms in the middle of winter. In summer the effect is reversed. Air cooled and moistened during its passage over the lake water is heated from below by the surrounding hot land, resulting, again, in convection and thunderstorms.

In coastal regions spring is typically a time of unstable weather when cold, moist air from the sea passes over land becoming warm under the spring sun. The Vancouver-Seattle area of the Pacific Northwest is affected in spring by air blowing off the North Pacific. Air is destabilized over the land, and the mountains provide additional lifting. The result is widespread cumulonimbus activity with showers and hail. This situation is reversed in autumn when the land is cooling and the air from the sea is still warm. The result is stratus and drizzle.

*Relative advection* is another mechanism with the potential to affect the stability of the atmosphere. In simple terms relative advection means that the air at surface is coming from one direction while the air aloft is coming from another. Therefore, layers of air of contrasting characteristics may be stacked on top of one another. In the central and southern U.S. in summer, warm, moist air from the Gulf of Mexico flows northward over the hot land and is destabilized. Being already hot and laden with moisture, it has the potential to produce violent thunderstorms. When cold air flows in aloft at heights on the order of 20,000 feet, the result is continuous instability from surface to great heights and this situation is what gives rise to the colossal storms and tornadoes that are well known in that part of the world.

## Stability and Moisture

Stability is a matter of degree. The air can be highly stable, highly unstable, or any gradation in between. Furthermore, the troposphere can contain layers of air of differing stability. Often these can be distinguished by the cumulus-type or stratus-type clouds forming in them. It is also worth mentioning that stability is a matter of relative rather than absolute temperatures: air can be warm and stable, cold and unstable, or vice versa. The air can also be dry or moist as well as stable or unstable.

Dry, stable air makes for easy flying and holds few surprises. The air is smooth and generally clear. Haze and smoke can, however, accumulate beneath an inversion and impede visibility.

Air that is unstable but so dry that no clouds will form in it usually is found only in desert areas. Thermal churning of the air gives a rough ride and causes dust devils, mentioned above.

Moisture in the air is what causes most of the visual pilot's problems. Moist, stable air produces stratus and fog, unless it is affected by frontal lifting in which case altostratus and nimbostratus may also occur.

A diurnal cycle, which is characteristic of moist, stable air will occur unless disturbed by the activity of weather systems. If nocturnal cooling brings the air to its dewpoint, fog and stratus form. After sunrise enough of the sun's energy will get through to heat the ground and cause these clouds to "burn off" during the day. The burning off is really evaporation of the clouds. We have seen that clouds, although often referred to as "vapor," are in fact composed of tiny droplets of water. These droplets can evaporate into true water vapor in which case the clouds will disappear.

In the early part of the day our world may be roofed with a sullen, gray overcast. However, as soon as the sun breaks through in just a few places to heat the ground directly, the clouds soon disappear. Remaining stratus may become stratocumulus; fogbanks may lift to become low decks of stratus or stratocumulus. At the end of the day, the sun's heat weakens and radiation from the earth once again exceeds insolation. The air cools once more and clouds form again.

In summer and early autumn, when the air is relatively dry, the hours of insolation are long and the midday sun is high in the sky. Clouds may not form until late in the night and may burn off soon after dawn. In winter, when the days are short and the midday sun is low in the sky, the air may clear briefly in the afternoon if at all.

What this means for the visual pilot is that in moist stable air, flying conditions will be at their best in the middle of the day. Unless active weather systems are present in the vicinity, pilots can safely delay departure until midmorning, confident of a fine afternoon, but they must be on the ground again before low clouds and fog can form in the evening. The later the clouds burn off, the earlier they will reform. This is something which the pilot can predict simply by looking at the sky and considering the time of day and the season of the year. A situation can arise in which the outer high clouds of a weather system cut off the sunlight so that the low clouds never do burn off. The weather system itself may then arrive and make its own contribution of clouds and rain. Early onset of a dark and cloudy night follows. Visual pilots are checkmated by this sequence of events and would be unwise to let frustration get the better of them.

A similar daily cycle takes place in the behavior of moist unstable air. In the early morning the sky may be clear. About midmorning cumulus clouds start to form, especially over hills, as insolation heats the ground and destabilizes the lower air. Two oppposing forces are at work here. The heating of the lower atmosphere, which tends toward destabilization and thermal lifting, also increases the air's capacity to hold water vapor without saturating. Whether clouds do or do not form, and whether they then increase or burn off, depends on which of these forces is dominant.

By midafternoon the day's potential for instability will be fully developed. The result may be cumulus, towering cumulus, or cumulonimbus clouds. These constitute the worst flying conditions of the day. Even so, visual pilots may find themselves flying in smooth, clear air over lowlands, wide valleys, lakes, or sea straits while various cumulus-type clouds sprout over hills, mountains, islands, or heated land surfaces.

As the ground cools in the evening convective activity dies away and the air becomes more stable. If strong thunderstorms have developed, however, especially if they are associated with a trough, front, or depression, they may continue into the night. Night-time cooling of air drenched by thundery rain is likely to produce fog or stratus. If convective activity is weak in the afternoon, VFR conditions probably will be good during the evening and night. If thunderstorms are still active toward dusk, however, the outlook is not good for night VFR.

The diurnal cycle of stability also affects the behavior of weather systems. The complexity of the weather as it actually occurs, rather than as described in the neat examples of the textbooks, lies in the interaction of many different processes.

By combining seasonal and diurnal stability effects, we can gain a useful picture of how the weather is likely to develop over the next 12 hours by little more than looking at the sky for evidence of moisture and stability. By considering the strength and direction of the wind and the signs given by mid- and upper-level clouds, the new pilot should, with study and practice, become able to predict the weather over the next few hours with at least enough certainty to be free of complete dependence on official sources of information, and well enough not to be caught by bad weather.

# 10

# Pressure Systems, Air Masses, and Fronts

## Sweeping lines on the map

**WE HAVE LOOKED** in some detail at temperature, water in the air, and atmospheric stability because they are features we can see and feel in progress around us, and they affect our short-range VFR operations hour by hour. The same is not true of the large-scale atmospheric pressure and circulation systems. They are elements of weather at the continental or oceanic scale, which may occupy the whole depth of the troposphere. Most of the activity in terms of clouds and precipitation takes place below 20,000 feet, but, even so, their analysis and prediction is a task of the most extreme complexity, which demands computer power as fast as the computer industry can produce it. Here we must regard these weather systems as the cards dealt to us, without really asking why. A proper treatment can be found in *Understanding Our Atmospheric Environment,* by Neiburger, Edinger, and Bonner, and *General Meteorology,* by Byers.

The structure of this chapter will follow a sequence of causes and effects that has at least an element of general validity. *Anticyclones* give rise to air masses. *Air masses* move out of their source regions and interact along lines called *fronts*. Swirls in major fronts produce *depressions*. Anticyclones, air masses, fronts, and depressions all produce weather phenomena affecting the VFR pilot.

## Barometric Pressure

Barometric pressure is charted by the aviation weather services, both for the earth's surface and for a series of pressure levels aloft, namely the 850-, 700-, 500-, 300-, 250-, and 150-millibar levels. These correspond roughly to 5,000, 10,000, 18,000, 30,000, 35,000, 39,000, and 45,000 feet above sea level. Surface charts map the distribution of pressure at surface. Upper air pressure charts map the elevations above sea level of the pressure surfaces they represent. The surface pressure chart is the one most important to us.

We might think that air would flow directly from an area of high pressure to one of lower pressure as soon as a pressure imbalance occurs. The earth's rotation, however, causes a phenomenon called the *Coriolis force,* which causes air to flow more nearly parallel to isobars than across them and which allows pressure imbalances to be sustained for long periods. The Coriolis force is not something we experience in our everyday lives, although it can be demonstrated experimentally. At this point, we have to take it on trust.

## Anticyclones

Anticyclones, or "highs," are large areas of above-average barometric pressure (1,020 to 1,025 millibars, up to extremes of 1,050 millibars). They are commonly on the order of 2,000 miles across, although of irregular outline. Their cores may shift slightly, but they tend to swell and shrink with the seasons rather than to move bodily from place to place.

Belts of high pressure occur around the earth between 20° and 40° North and South latitude, and areas of high pressure form over the Arctic and Antarctic. The uneven distribution of land and sea causes the subtropical high-pressure belts to be broken up into cells. Two warm highs affect North America: the Hawaiian and the Azores highs, located, respectively, over the Pacific and Atlantic Oceans. A cold high develops over the Arctic in the winter months. Unless we live in the core of a major high-pressure area, and not many of us do, we will be affected by a high spreading over us or shrinking away from us and by wind circulation around it rather than by its passage.

Anticyclones are areas of air sinking down through the troposphere and diverging at surface. The air becomes warmer as it subsides, because of adiabatic compression. As it becomes warmer, its relative humidity decreases. Typical sink rates are on the order of 100 feet per day from 20,000 feet down to 5,000 feet. Within an anticyclone, pressure gradients are weak, resulting in light winds. Because air masses do not come into conflict within them, fronts cannot occur within anticyclones. The sinking air prevents instability from developing to any great depth; the deep instability requisite for thunderstorm development is therefore absent.

Anticyclones are areas of settled weather. Nevertheless low stratus and fog can present a serious and persistent problem for visual pilots. Once a cloud deck forms, air subsiding above it is being warmed by adiabatic compression and may become warmer than the air in the cloud deck. This produces a strong inversion, which may persist for as long as the anticyclone remains in place. At night the inversion is reinforced by radiation from the top of the cloud. The result is a persistent deck of stratus or stratocumulus known as "anticyclonic gloom." Domestic and industrial pollution trapped beneath the inversion accumulates in the still air to make this situation worse.

Convection can occur in warm anticyclones. In temperate latitudes, 20% of anticyclones limit convection to altitudes less than 3,000 feet above sea level; 65% confine it to less than 6,000 feet; 15% allow some convection to heights above 6,000 feet.

Light winds blow around anticyclones in a clockwise direction in the northern hemisphere ("hIGH on the rIGHt" seems to jog the memory). Thermally induced winds can be quite strong in the afternoon, but any overall increase in wind speed, especially if continued through the late evening and morning, is an indication that the high may be weakening and that more changeable weather is on the way.

The upper sky in an anticyclone is clear except for wisps of cirrus. If the cirrus becomes organized into bands across the sky, it signifies the presence of a jetstream aloft and a possible end to the fine weather over the next few days. Persistent thickening of the cirrus, especially with the appearance of other cloud types, signifies the approach of a weather system from outside the anticyclone.

Anticyclonic weather, although quiet, is not always benign to the pilot, instrument or visual. Conditions are ideal for the formation and persistence of fog and low stratus or stratocumulus. These can checkmate the pilot for days or even weeks on end.

How, then, do we recognize a high-pressure system? The best visual indication is an upper sky free of clouds. If we are covered by unbroken low clouds or fog, we cannot see the upper sky, but sometimes there are breaks through which we can see the clear blue sky above. Light winds are another good indicator. Absence of precipitation, in combination with other indicators, also might suggest an anticyclone in place. Most obvious of all is the high barometric pressure, or the altimeter setting, which is the same thing. The earth's average atmospheric pressure at surface is 1,013 millibars, or 29.92 inches of mercury. We can see, then, that altimeter settings higher than 30.00 give a fair indication of anticyclonic conditions.

## Air Masses

When air is conditioned as to temperature and moisture content by prolonged contact with an extensive and uniform part of the earth's surface, it acquires characteristics which are the same over wide areas. The result is known as an *air mass*. The area over which the conditioning takes place is known as a *source area*. A cold continent or a warm ocean are typical source areas.

The conditions needed for air mass formation occur most readily in large areas of high pressure. Accordingly, the source regions affecting North America are the arctic and subtropical anticyclones. These source regions give rise to five main air mass types: Arctic, Maritime Polar, Continental Polar, Maritime Tropical, and Continental Tropical. The Arctic and Polar air masses affect the continent from the north and are cold. The Tropical air masses affect the continent from its interior, from the southeast and southwest, and are warm. The Continental air masses are dry. The Maritime air masses are moist. An air mass moving over a warmer surface, whether land or water, becomes unstable. An air mass moving over a colder surface becomes stable.

With these basic facts in mind, we can see the causes of a number of weather phenomena well known to inhabitants of the areas affected:

1. From time to time in winter the large-scale distribution of pressure around the Great Lakes is such that a stream of Continental Polar air is drawn from its source region in northern Canada and moves southward across the Great Lakes. The results have been mentioned in the chapter on atmospheric stability. On crossing the relatively warm Great Lakes, the air mass is warmed from below, moistened, and destabilized, producing heavy snow showers and even thunderstorms on the south side of the Lakes.

2. Along the south coast of Newfoundland, warm, moist Maritime Tropical air moving northward over the cold Labrador Current is stabilized and cooled, causing extensive fog.

3. When Maritime Tropical air, already warm and laden with moisture, flows up the Mississippi Valley in summer, it passes over a land surface hotter than its source region over the Gulf of Mexico. It becomes unstable and, because of the heat and moisture available, causes violent thunderstorms.

Movement of airstreams can be seen most clearly through the medium of time-lapse satellite photography, which is shown on some television weather programs. We can see cloud forming and dispersing, air swirling around depressions, and fronts forming, moving, and dissipating. By knowing the major circulation pattern affecting their area, and bearing in mind the time of year, where the air is coming from, and how it is being influenced by the surfaces over which it is passing, pilots can gain at least some idea of what weather to expect and how long it may last.

How do we recognize "air-mass weather?" The regional wind remains constant. This does not, however, guarantee constant weather conditions during the day, because of the diurnal cycles of temperature and stability discussed in Chapter 9. Air-mass weather may be difficult to recognize without recourse to an indication (newspaper, television, radio) of the layout of the wind circulation pattern on the continental scale. Professional meteorologists with local experience may be able to look at the sky and recognize "air-mass weather," but we are not likely to be in that position. Even so, a clear upper sky with a steady wind throughout the day and a steady altimeter setting between 29.50 and 30.00 could suggest air mass weather.

Because the flow of air is caused by large pressure systems, and the air is flowing around or between them, it follows that barometric pressure will be neither particularly high nor particularly low. The upper sky is clear because the large-scale upward motion of air associated with fronts and depressions is absent. The clouds forming will be those resulting from the temperature and moisture content of the lower 5,000 feet of the atmosphere and its interaction with the land and water surface over which it is passing. As this interaction takes place within the bottom mile of the sky, clouds will be initiated in that air. The word "initiated" is used on purpose because air-mass thunderstorms and cumulonimbus cells initiated below 5,000 feet can extend to great heights and cause other clouds to form at those heights.

## Fronts

Air masses of differing characteristics do not mix readily. Instead, when they meet, a linear discontinuity, called a front, arises. Along a front the warmer air is wedged up over the colder air. If the air is moist enough, clouds and precipitation form just as surely as if the air had been forced to rise over a range of hills. The analogy is not exact because the slope of the front is much less abrupt than that of hills and extends to much greater heights.

One of the main characteristics of a front is that it moves in a direction perpendicular to itself. Sometimes the movement is very slow and the front may even stop or reverse. The movement of such a front is hard to predict.

Fronts occupy a prominent position in the study of meteorology and pilots' perception of the weather. The textbooks and courses describe warm, cold, and occluded fronts. Some go into more detail about cold and warm occlusions, frontal cloud types in moist and dry, stable and unstable air, ice accretion, and other matters. Occasionally we see a classic front behaving exactly as described. More often we do not; we know only that there was a period of bad weather. It is even possible for a front to pass over us without our being aware of it because no obvious weather conditions occurred. We are mystified by the discrepancy between theory and practice. The reason is that the whole story is far more complicated than what we have been told.

The classic warm or cold front is of only one type: the *ana-front*. There are also *kata-fronts,* which are weak fronts with only slight weather activity along them and only low- and mid-level clouds, not the full assemblage of higher clouds.

The terrain over which a front is passing will affect it, as will the characteristics of the air masses interacting along it. If the front is moving so slowly that its full range of effects takes more than 24 hours to pass over a place on the ground, diurnal effects on the weather produced at that point may be noticeable. The full complexity of fronts is almost without limit.

A large part of the problem is that we try to impose rigid definitions and chart symbols on weather systems that are both variable and amorphous. It may be a matter of opinion whether or not a front exists at a particular time and place. Cases are far from rare when two fronts of the same type follow one another closely, separating air masses which are, for example, cold, colder, and coldest.

In many cases cold and warm fronts cannot be clearly differentiated. A front may behave as one or the other at different points along its length, which may range from a few hundred miles to several thousand. It may exist only aloft, in which case it is called an *upper front.* Its character will vary along its length according to the landforms and water surfaces over which it passes, the passage of time, diurnal temperature effects, and the character of the air masses involved. Fronts can fade out. They can appear where none are reported. Two of a kind can follow in quick succession. No two fronts are alike. "Troughs" and "waves" can behave like fronts.

The summary of all this is surprisingly simple. There is a group of weather features existing along straight or gently curving lines. They advance perpendicular to themselves. Sometimes they stand still; sometimes they reverse. These linear features are called fronts, occlusions, waves, troughs, frontal systems, trowals, upper fronts, and various other names. Waves and troughs are not technically fronts, but in some cases the weather they produce is similar to that of a certain type of front. We will take "front" as a generic term. A high proportion of these features will produce weather at some time and place that the VFR pilot will be unable to penetrate successfully, especially at night, in winter, or in mountainous country.

Small local fronts can occur wherever air of dissimilar properties is in contact. An example is the *sea breeze front*. Such fronts are generally of little significance; nevertheless, when conditions are right, they can touch off thunderstorms.

The weather associated with a front can be classified into three zones: prefrontal, frontal, and postfrontal. The prefrontal zone is a belt of weather that will be seen to deteriorate from the point of view of a ground observer. Frontal weather is a belt of the worst weather the front will product. Postfrontal weather is generally a belt of improving weather following behind the front, although in some cases the improvement may not be great and may be offset, from the VFR pilot's point of view, by nightfall. The zones range in width from 50 to 250 miles perpendicular to the front.

### AN EXAMPLE

Let us examine a front as described in the aviation weather reports. We will follow it through from the area forecast to its predicted effects on one place, as seen in the terminal forecasts, and then to some surface actuals to see what actually happened. This is a front coming ashore from the North Pacific in March, and we will review its forecast and actual effects on Vancouver,

British Columbia. We are contemplating a flight in the middle of the day.

The prognosis section of the area forecast reads:

> Maritime front in a line from just west of Port Hardy to just west of Victoria at 10 a.m. (local time), moving to lie in a line from 60 miles south of Prince Rupert to Spokane at 10 p.m. (local time).

This front is about 700 miles long with its north end in a 994-millibar low of irregular shape but something like 1,000 miles across. The isobars on the chart do not resemble the straight or smoothly curved lines of the textbooks. The front already will be affecting the Vancouver-Victoria area by 10 a.m., and frontal passage may be expected within the next few hours. The situation is shown in Figure 10.1

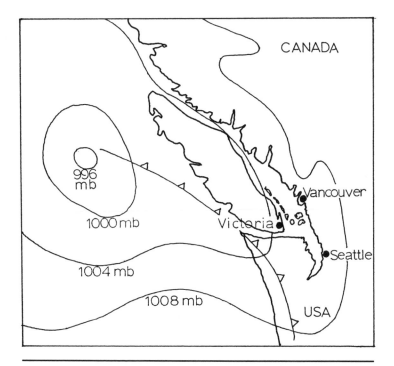

**Fig. 10.1. Surface synoptic chart showing situation described above (0400 local time, March 13, 1986).**

The effects on the area forecast region covering the southern coast of British Columbia are predicted to be as follows:

> Vicinity to 120 miles northeast of the front: 3,000 to 4,000 scattered, occasionally broken, tops at 8,000. 8,000 to 10,000 broken, variable overcast, layers to 18,000, light rain showers, occasional visibilities of 2 to 6 miles in light rain showers and fog, except 1 to 4 miles visibility in light snow showers at the higher levels. Occasional ceilings 200 to 1,000 feet above ground and visibilities ½ to 4 miles in precipitation and fog. Precipitation generally light rain along the west coast of Vancouver Island. Becoming, southwest of the front, 3,000 to 4,000 broken, occasionally scattered, cumulus, tops at 8,000, with 3,000 scattered, occasionally broken, towering cumulus, tops at 18,000, and light rain showers. Isolated cumulonimbus to 20,000 with visibilities 1 to 3 miles in rain showers and fog, risk of light thundershowers. Precipitation giving visibilities of 1 to 4 miles in light snow showers at the higher levels. Few stratus ceilings of 500 to 1,500 feet above ground. Surface winds southeasterly gusting 35 knots over northern Vancouver Island, abating by the middle of the period.

In this case the frontal zone itself is probably quite narrow, so the weather is divided into prefrontal and postfrontal weather only. This region is one of extremely varied terrain, from sea straits, islands, and coastal lowlands to 10,000-foot mountains. Every possible sky condition is forecast, from quite reasonable VFR down to conditions below IFR minima at many airports with instrument approaches. So what is forecast for our actual area of operations?

The Vancouver terminal forecast, valid from 9 a.m. (local time) today until 9 a.m. tomorrow, reads:

> 4,000 scattered, ceiling 10,000 broken, occasional ceilings 4,000 broken, 8,000 overcast, few light rain showers. From 11 a.m. (local), 3,000 scattered, ceiling 5,000 overcast, light rain showers, occasional ceilings 3,000 overcast, visibility 3 miles in light rain showers and fog. From 3 p.m. (local), ceiling 3,000 broken, towering cumulus, light rain showers, occasionally 3,000 scattered.

In the middle of the morning we have good VFR conditions. There are some clouds, probably on or near the hills, at 4,000 feet, with a ceiling at 10,000 feet and blue sky visible above. The clouds will lower and thicken during the morning. Even so, the worst conditions forecast between 9 a.m. and 11 a.m. will not trouble us and probably will be isolated showers, which we can avoid. Frontal passage is expected around 3 p.m., with 3,000-foot ceilings and towering cumulus with their accompanying showers. Nevertheless, no winds are forecast above 10 knots, and no visibilities less than 6 miles, except in showers. We would be flying in a quiet, gray sky with clouds on the mountains and isolated showers.

By following the surface actuals for the day, we can see how things turned out. At 10 a.m. we were in the prefrontal weather and Vancouver reported:

> 8,000 scattered, estimated ceiling 12,000 broken, 22,000 broken, visibility 20 miles, barometric pressure 1004.8 millibars, temperature 8°C, dewpoint 2°C, wind from the east at 4 knots, altimeter setting 29.67.

The cloud types reported were altocumulus, altostratus, and cirrus. The weather was actually brighter than forecast, and very quiet. All through the middle of the day there was scattered cumulus based between 2,000 and 3,000 feet. An altocumulus layer lowered and thickened to become a broken stratocumulus ceiling based between 4,000 and 6,000 feet. Visibility continued between 15 and 25 miles in light rain showers.

By 3 p.m. the cloud layer at 5,500 feet had thinned out, leaving a higher ceiling at 11,000 feet, as before. The showers had ceased. As frontal passage had been forecast for about this time, we might have thought that this had actually occurred. Toward dusk, however, the cloud layer at 5,500 feet thickened again to form a ceiling, and showers began again. Close to 8 p.m. the front passed, with embedded towering cumulus giving 2,000-foot ceilings and visibilities of 4 miles in rain showers and fog, with scattered cloud at 800 feet. This would not have been a good situation to have run into while flying VFR at night. After about 2 hours the ceiling lifted but visibility remained at about 5–8 miles in rain showers.

If we had believed that the front had passed, arrival of the bad weather after dark could have caught us on the wrong foot. Any worsening of the weather around nightfall is a definite danger signal to the visual pilot. The sequence of measured weather features—wind speed, wind direction, temperature, and barometric pressure—all showed very slight changes, and even these did not occur in the manner that might have been expected.

A trained meteorologist can explain what happened by reading the sequence after the event, but this is very different from the situation facing the new visual pilot as he tries to unravel what is going on as it actually happens. Nevertheless, we have to base our plans on the weather as forecast. If it turns out better, we are in luck. If we find that it is worse than forecast, we should at once become suspicious and base our immediate actions on what we can see around us.

### GENERALIZATIONS

A few generalizations about fronts hold good:

1. Although the classic cold front arrives more abruptly and produces a shorter period of worse weather than the classic warm front, any one type of front cannot be said to produce inherently worse weather than another.

2. Although weather typically improves after a front has passed, the improvement does not always bring about VFR conditions. If the front passes late in the day, nightfall and its effects may negate any postfrontal improvement in the weather.

3. The faster a front moves, the more violent the weather it will produce. We may think of typical speeds of movement as being on the order of 20–30 knots, but any speed is possible, from no movement at all to extremes of 50–60 knots.

4. The more slowly a front moves, the more difficult its movement is to predict.

5. The more rugged the terrain over which a front passes, the worse and more unpredictable the weather it will produce.

6. Fronts move faster and produce worse weather in winter than in summer.

7. Fronts do not usually move faster than 15%–25% of the cruising speed of a light aircraft.

## FRONTAL SLOPES

Typical slopes of frontal surfaces are 1 in 100 to 1 in 300 for a warm front and 1 in 50 to 1 in 150 for a cold front. This has several interesting implications. Because fronts are shown on weather charts as lines and because textbook cross-sections through fronts have greatly exaggerated vertical scales, we tend to think of a front as a steeply inclined or vertical wall. This is far from true. A warm front is a gently sloping surface, no steeper than a railroad grade. If the frontal slope is 1 in 200, the frontal surface will be only 5,000 feet above ground at a position 200 miles ahead of the surface line of the front. Indeed, the surface "line" is more likely to be a zone of partly mixed air some 10–15 miles across.

Clouds form in the warm air overrunning the frontal surface. Therefore, in looking at the base of a frontal cloud layer, such as altostratus, we actually are looking at the frontal surface that defines the base of the cloud and is the reason for its existence. Moreover, if a warm front has a slope of 1 in 200 and is moving at 20 knots, it follows that the cloud base will lower over a place on the ground at about 500 feet per hour.

A pilot flying toward such a front at a groundspeed of 100 knots will encounter cloud bases that sink lower by 3,000 feet for each hour that he flies toward the front. Closer to the front, clouds may be forming in the underlying cold air as a result of saturation by rain falling from above the frontal surface. Of course, fronts move at various speeds and have various slopes, but the concept is worth thinking about.

Because cold frontal slopes tend to be steeper than warm frontal slopes, the surface of a cold front may be 5,000 feet above ground only 50 miles behind the surface line of the front. In fact, if the front close to the ground is steepening and overturning, as a result of ground friction, this distance may be even less. Because of the relatively steep frontal slope and more pronounced lifting, some cold fronts actually do appear as walls of towering cumulus and cumulonimbus when viewed both from ahead of and behind the front.

If a front has the leading-edge appearance of a warm front and the trailing-edge appearance of a cold front, which is it? It may be an occluded front, but let us leave this subject because each front is its own special case.

## DIRECTIONS OF FLIGHT RELATIVE TO A FRONT

In relation to a front, there are three possible directions of flight: toward, away from, or parallel to. These directions of flight can be ahead of the front or behind it.

Ahead of a front, flying toward it with the idea of beating it to the intended destination is unwise unless the pilot is sure of being on the ground at least 4 or 5 hours before the front's estimated time of arrival. The zone of worst weather may extend 100 miles ahead of the surface line of the front. This zone can easily take 5–10 hours to pass a place on the ground with the result that unflyable weather can begin several hours ahead of the estimated time of arrival of the front. Sometimes small waves pass along fronts, causing a repetition of the frontal weather before it is supposed to arrive or after it is supposed to have passed.

Although we know that air is compressible, it acts as an incompressible fluid at the small pressure differences occurring in the atmosphere, and indeed around the wings of light aircraft. Therefore, as an example, we might see clouds caused by mountains extending several miles upwind of the mountains causing them. Correspondingly, a range of hills will give added lift to a frontal surface passing over them and may cause clouds to form in the air moving ahead of the front. If a range of hills lies in the path of a front, clouds and precipitation will develop over them more and sooner than over neighboring flatlands. Frontal thunderstorms form preferentially over hills. Weather forecasts can predict the behavior of a front in general but seldom in detail.

Flying toward the trailing edge of a front is harmless, providing that the postfrontal weather is flyable, bearing in mind that the front moves on its own good time. A light aircraft will overtake the front and therefore will enter deteriorating weather even though the weather as seen from a place on the ground may be improving. Just as hills will cause foul weather farther ahead of the front than do flatlands, so the effects of the front will linger in the hills longer after it has passed. This is especially true in mountains high and steep enough to cast shadows that interfere with solar heating and drying of air saturated by rain. Flying toward a front under VFR should be carried out with some line of retreat in mind.

A flight parallel to the leading edge of the front can hold

unpleasant surprises because of the front's uneven advance. When flying in mountains, a pilot must have some route by which to break away from the front altogether, to avoid the risk of being trapped. Flying parallel to the trailing edge of a front is likely to succeed because the weather along the route should be improving.

Flight away from the leading edge of a front will succeed provided that the pilot can make an orderly departure before the weather deteriorates too much. He should have an alternate airport in mind not too far along his route in case he should need to land in the early stages of his flight and his departure airport has been weathered in. Flight away from the trailing edge of a front is likely to meet only improving weather.

## PREFLIGHT INFORMATION NEEDED

The visual pilot planning a flight near a front needs the following information:

1. Present and predicted positions of the front over the next 4–12 hours, depending on the operations intended.

2. Expected time of frontal passage over the intended area of operations.

3. The prefrontal weather.

4. The worst conditions to be expected of the front over the intended flight track.

5. The postfrontal weather.

# Depressions

North America is affected by three major frontal zones: the Continental Arctic front, the Maritime Arctic front, and the Maritime Polar front. They are the boundaries between air masses that are so different they will not mix. Sometimes these fronts are inactive and the weather map may show isolated and discontinuous fronts or none at all. At other times the activity is such that the charts show festoons of continuous lines curving across the whole continent.

It is a fact of physics that whenever a fluid shears against another fluid, or against a solid surface, waves form. This is true for the water shearing over a riverbed, air blowing over water, and air masses shearing along fronts.

From time to time, waves form along active fronts and may intensify as they move eastward along the front. The wave then may "break" like a breaking surf to form a system of closed wind circulation, with the lowest barometric pressure at its core, called a "depression," "cyclone," or "low." (The word "cyclone" sometimes is used to indicate a tropical cyclone, which is the same as a typhoon or hurricane.) A fully developed frontal depression may be 1,000–2,000 miles across.

Whereas an anticyclone slowly expands and contracts about its core area or, at most, small, closed high-pressure cells may become detached and wander about, a fully formed depression maintains a roughly constant size but its core moves continually. Stationary depressions are the exception rather than the rule.

Small secondary depressions can form along the fronts radiating from a depression. These can form suddenly, are difficult to detect because of their small size and rapid formation, and can produce vicious weather.

Some areas favor the formation of depressions and, as a result, the barometric pressure, averaged seasonally, is lower there than elsewhere. Examples are the Aleutian Islands and the Denmark Strait between Greenland and Iceland. The atmospheric features resulting are known as the Aleutian and Icelandic Lows. They are not permanent depressions in the sense that the Hawaiian and Azores Highs are permanent anticyclones.

The main areas in which frontal depressions form around the North American continent are the northeastern Pacific, the northwestern Atlantic, and the southeastern U.S. Typically, frontal depressions forming over the Denmark Strait reach maturity over the eastern Atlantic and reach the fully occluded stage over the British Isles and northwest Europe. Over eastern North America, formation over the southeastern states proceeds to occlusion over the Maritime Provinces of Canada. Along an active major front in the northern hemisphere may be a chain of depressions moving northeastward. The most northeasterly may be occluding already; its successor, perhaps 1,000 miles behind, may be fully mature, and 1,000 miles behind it a new wave may be forming on the frontal line.

Brisk to strong winds circulate around a depression in an anticlockwise direction (in the northern hemisphere). A direct correlation exists between the lowness of pressure at the core of a depression and the speed of the circulating winds, the speed at which the system moves over the earth's surface, and the foulness of the associated weather. "Low on the Left" is an aid to the memory.

In the northern hemisphere the fronts radiate from the righthand side of the core of a depression with respect to its direction of movement. The lefthand side is quieter with less rapidly changing conditions, although this may equate to several days of continuous low clouds, rain, drizzle, or snow. Depression tracks are recorded by the meteorological services and are charted in some publications.

Satellite photographs show spiraling bands of matted clouds around the core of a depression, with fronts visible as tails of clouds radiating from it. The depression is an area in which air is spiraling inward (converging) and upward toward the tropopause. Closed lows reaching the 500-millibar (18,000-foot) level are, however, relatively uncommon.

The overall conditions are the exact opposite of those in an anticyclone. The whole tendency is toward instability and the formation of clouds to great heights. When seen from the ground, the outer edges of a depression appear as continuous mid- and high-level clouds. Conditions worsen toward the core of the low, whether the observer is moving toward it or it is moving toward him, or some combination of the two.

Depressions shown on the weather charts and reported in area forecasts are tagged with the barometric pressure at their cores—e.g., "a 990-millibar low." Any self-contained low with a surface core pressure lower than 995 millibars is strongly active. Normal, but violent, depressions have been observed with core pressures as low as 940 millibars.

If a depression is 1,000 miles across and moving at 30 knots, we can see that a cross-country flight of 2 hours and 200 miles is small in relation to the size and lifespan of the depression.

If a depression is affecting the area of intended flight, the pilot needs the answers to these questions:

1. Where is it?
2. How intense is it?
3. Is it intensifying or weakening?
4. How is it moving?
5. What weather is it producing?

## NONFRONTAL DEPRESSIONS

Nonfrontal depressions also occur, and they tend to be smaller than the full-fledged frontal types with one important exception. Largest of the nonfrontal depressions is the *hurricane.* Enough has been written about them for us to omit further mention here. They are sufficiently obvious that the lightplane pilot is unlikely to become entangled with one accidentally.

Other varieties of nonfrontal depressions are smaller in size and of shorter duration than the hurricane. *Polar lows* form in cold, unstable air. Typical diameters are on the order of 100 miles. Sometimes a cold front extends to the west and produces rain or snow showers. *Thermal lows* form over a hot landmass in summer. *Thundery lows* form in troughs of low pressure with suitable upper wind conditions. Best known in central Europe and northeastern Australia, they cause days of thundery weather. *Lee depressions* form downwind of mountain ranges in conditions of a suitable regional circulation. They are well known in the lee of the U.S. Rockies, from which they move eastward across the Great Plains.

We all know the feeling when a fine day with clear skies, settled weather, and a good forecast suddenly comes unstuck, with such foul weather that we can hardly believe what we see. Often a small secondary or nonfrontal depression or trough is the culprit. Needless to say, sudden deteriorations like these can endanger visual pilots. When they see the weather turning against them, it is time for a serious reappraisal of whatever flying they had in mind.

## Use of the Altimeter

For a direct, no-nonsense forecast, let us look at the face of a household barometer. It is marked "Storm, Rain, Change, Fair, Very Dry." These words are directly related to certain barometric pressures. The midpoint in the middle of "Change" is 29.50 inches of mercury. The subdivisions do not always hold true, but long before the science of meteorology came into being in its modern form, people recognized a direct connection between barometric pressure, especially pressure trends, and the forthcoming weather.

The aircraft's altimeter is essentially an aneroid barometer. The altimeter settings in the Kollsman window are barometric pressures in inches of mercury.

Altimeter settings are broadcast on ATIS frequencies. The pilot hears them being given to other pilots, and often is given a setting without asking. If pilots are flying out of a field with no facilities whatever, at least they should know the field elevation so they can set the altimeter to a good approximation accordingly. Even if the altimeter is out of calibration by 100 feet, that represents an error of only 0.10 inches of mercury.

The pilot almost certainly will know when the aircraft was last flown. It is fair to assume that the altimeter was set then. When he resets it, he can see by how much the setting has changed. If the aircraft last flew more than 24 hours ago, the change is of limited relevance. But a rapidly falling altimeter setting is a certain warning of foul weather and not to be disregarded. Occasionally, if a situation is developing faster than the aviation weather services can analyze it, the altimeter setting may be the only warning available.

As an example, a fall in altimeter setting of 0.10 inches of mercury in an hour might forecast the approach of a vigorous and fast-moving depression. That is the pressure fall at a place on the ground in the path of the depression. If a pilot were to be flying toward the depression, he would be flying over a succession of ground stations, each of which would be experiencing falling barometric pressure. An airport might, for instance, transmit an altimeter setting 0.20 inches of mercury lower than the setting the pilot received half an hour earlier from an airport 50 miles back along the route. Lowering altimeter settings do not occur in isolation. The view ahead will be encumbered by lower, darker, and thicker cloud, or in some cases by cumulonimbus anvils, which can be seen from great distances. Once alerted to deteriorating weather, the pilot can decide how bad to let things get before having to turn around.

The breakeven point between good and bad weather is roughly the 30.00 altimeter setting. The ICAO Standard pressure is 29.92 inches of mercury, which is close. If the altimeter setting begins with "3," the weather is likely to be settled. (It may be settled and zero-zero in fog, too, but there will not be too many surprises.) If the setting begins with "2," more changeable weather can be expected, possibly with wind, low clouds, and rain. Good weather often occurs with altimeter settings between 29.50 and 30.00, but the generalization is usable. All this good information is flying around free, so why not use it?

Pressure systems, air masses, and fronts are large-scale features of global weather and climate. Although we cannot expect to acquire a detailed understanding of them without an advanced study of meteorology, an elementary knowledge of their causes and effects is important to enable us to interpret the weather we see around us and in which we fly—or sometimes do not fly as the case may be.

# 11

# Day, Night, Seasons, and Terrain: Visibility Effects

## What the eye doesn't see

**HAVING DISCUSSED** the importance of outside visual references and the weather phenomena that can obscure them, we now come to some features affecting visibility which change either regularly or not at all. Night and day are predictable; differences in visibility between them are obvious. Night introduces specific illusions against which pilots must guard themselves. Weather phenomena and the accuracy with which they are reported vary between night and day. The changing seasons bring with them, and indeed are characterized by, differing types of weather. Landforms and water masses may have their own characteristic weather.

## Effects of Night and Day

The main effect of night is that it conceals a substantial portion of the available attitude references. Those remaining are sometimes illusory. "Night" itself can encompass a wide range of visibility conditions.

## HORIZON REFERENCES

The light of a full moon in a clear sky can provide visibility almost as good as daylight. Snow-covered mountains, for example, can be seen 50 miles away in clear air under a full moon. By contrast, over open water or uninhabited land on a moonless night beneath a thick overcast, horizon references may be absent. Moonlight and haze can have the same effect. In settled country, lights on the ground often provide a completely adequate horizon reference even without light from the moon or stars. If ground lights are few and far between, however, they can provide a misleading horizon reference. New pilots, therefore, must plan their night flights considering the availability of horizon references.

Pilots are deprived of horizon references between the moment of takeoff and the time when they are high enough to gain a usable view of lights on the ground. Runway lights and the distant horizon may be obscured by the aircraft's structure, and the departure area may be devoid of ground lights. Control of the aircraft by instruments is essential. Serious accidents have befallen even experienced pilots who have flown large aircraft back into the ground or water, or hit obstructions for no reason other than failure to establish precise control by instruments immediately after a night takeoff. Above 500 feet, in VFR conditions, the pilot should be able to see enough ground detail to resume visual flight.

The new pilot must realize that in some conditions attitude references fade with the onset of night. Either the sky must provide enough light to illuminate the ground or enough dwellings must be present on the ground to light up at dusk and provide a substitute. VFR on top, especially with high cloud above, is another situation in which good horizon references by day will disappear by night. If ground lights are to be used for attitude reference, there must be enough forward visibility for a good spread to be visible. Flight in marginal VFR conditions is obviously risky in this respect. Extended flight in marginal VFR conditions is hardly to be recommended by day, and certainly not by night.

Cross-country night VFR in anything other than settled fine weather is hardly to be recommended to the new pilot because of the risk of flying into clouds inadvertently, getting lost, or hitting obstructions. Some countries do not allow night VFR at all. Others do not allow night VFR by air carriers. That should tell us something.

## NIGHT ILLUSIONS

Visual illusions at night begin as soon as the aircraft moves from its tiedown. Poorly marked obstructions may be a hazard at small airports. A patch of darkness between the lights may turn out to be a parked aircraft. This seems to be especially true if the aircraft has a small, solid shape with booms extending from it, such as a Cessna 337 or a helicopter.

Night illusions come out and bite on the approach to land. The most common illusion causes the pilot to fly too low on final approach. The use of VASIS (Visual Approach Slope Indicator System), if available, will help to guard against this. This is especially prevalent if the airport is higher than surrounding lighted areas. Towns that have curved streets or those that are built on slopes can provide false horizon references. If the relationship between runway length and width is the same as another runway with which the pilot is familiar, but the runway he is approaching is both shorter and narrower, he will believe that he is farther back from the runway and higher than he actually is.

When approaching to land in a crosswind, the pilot may set up a sideslip appropriate to what he can see of the approach lights. Low down over the runway the crosswind may be weaker, and the dark area of a wide runway may not show the pilot that he is sideslipping into a wind that is no longer there. He then may touch down with enough drift to cause a rough landing in a tricycle-gear aircraft, or a groundloop in a taildragger. Any of these illusions can be worsened by poor visibility, especially with rain streaming off the windshield.

## CLOUDS

At night the lower atmosphere loses heat to the earth's surface, which is cooling by radiating heat into space. As the air temperature cools toward the dewpoint (i.e., the temperature-dewpoint spread closes), less lifting is needed to produce condensation, with the result that clouds become lower, thicker, and more continuous. If the temperature cools to the dewpoint itself, no lifting will be needed at all and clouds will form spontaneously.

Nightfall thus tends to bring with it an increase in cloudiness. These clouds are less easy to see, either by people on the ground or by pilots in flight. The reporting services are also less active at night. Risk of inadvertent entry into clouds therefore is much greater by night than by day. If outside visual references are sparse, the pilot may not realize immediately that the plane has entered clouds.

We have seen that, if control of the aircraft is lost in clouds, there may not be enough space between the clouds and the ground for control to be regained, assuming that the aircraft does not break up in flight because of its uncontrolled maneuvers. At night, even if it emerges into clear air, ground references may be insufficient or misleading, thereby reducing the pilot's already slim chances of survival.

We have looked at the tendency toward atmospheric stability at night. Convective activity, unless strongly developed, tends to die away in the evening. A thundery afternoon may be followed by a clear, calm night. However, if moist, unstable air moves over large bodies of water, whether in air mass conditions or because of the land breeze effect, the water, being warmer at night than the land, can touch off thunderstorms. Both air mass and frontal thunderstorms can continue to form at night. Even so, thunderstorms rarely are as concentrated or as intense at night as by day.

**THE MOON**

The phases of the moon are absolutely predictable and are published in some newspapers, almanacs, and similar material. A half-moon or fuller provides good night VFR conditions when total cloud cover is less than five-tenths, that is, when no cloud layer is reported as more than "scattered."

### ALERTNESS

Humans are daylight animals. Some of us are more alert in the morning than in the evening, or vice versa. Total sleep requirements and responses to lack of sleep vary. Very few of us, however, are really alert between midnight and 6 a.m., and it is widely recognized that all of the body's daily cycles hit a low point between 3 a.m. and 5 a.m. We tend to become more alert at first light than we are in the predawn darkness. To compound matters, aviation services are at a minimum between midnight and 6 a.m.

At nightfall the visual pilot's need for outside visual references by which to navigate and maintain control of the aircraft becomes more acute as some of these references disappear in the darkness and others become misleading. At the same time, the lower atmosphere cools and water vapor condenses. While this is going on, the pilot's alertness is on the way down and aviation ground facilities progressively close. Pilots' needs increase while, at the same time, the means to satisfy those needs decrease. To lengthen the odds in their favor, new pilots need significantly better weather for night VFR than for day VFR. Although the regulations make no distinction between day and night VFR in terms of weather minima, and a very small one in terms of required training, the new pilot should undertake night VFR with caution.

## Seasonal Effects

The world is divided into many climatic zones. They are distinguished, among other things, by the character and duration of the four commonly acknowledged seasons. Continental North America has six or seven distinctly different climates. These are, however, only broad zones of roughly similar climatic conditions, and the boundaries between them are drawn somewhat arbitrarily. Each zone can be further divided into a mosaic of local climates and microclimates. Even the concept of four seasons may itself be arbitrary.

The coastal Pacific Northwest, for example, has eight clearly discernible seasons: winter, false spring, spring, early summer, the rainy interlude, high summer, Indian summer, and fall. From year to year these are shorter or longer and their characteristic weather forms are more or less intense. They do occur, however, and are recognizable to the cunning pilot. Other areas have their own arrangement of seasons and microseasons.

Winter brings longer nights, colder temperatures, denser and thicker cloud cover, stronger winds, snow and ice in some regions, increased rain in others. Pressure systems and fronts tend to develop more strongly and move faster. All of these add up to poor VFR flying conditions, demanding more skill on the part of the pilot in interpreting the sky.

With spring comes rapidly changing, and sometimes rough, weather as the land heats up while the air is still cold from winter. In some areas melting snow readily saturates the lower air and causes fog.

Summer is characterized by generally more agreeable flying conditions. It is, however, the thunderstorm season, and convective weather features of all kinds are more intense. Higher air temperatures allow the air to hold more water vapor, and the results are apt to be more violent when it condenses. The fog season is summer in some parts of North America, winter in others. Summer in the West is the time to be on guard against the effects of high density altitudes.

Fall is a time of cooling and moistening. Air still warm from summer and holding plenty of moisture becomes stable over the cooling land, and its contained moisture condenses to form extensive cloud and fog. Wind speeds begin to increase.

The figures accompanying Chapter 7 show the seasonal distribution of IFR conditions, fog, and thunderstorms.

Stability of the atmosphere goes through a seasonal cycle. Stable conditions are typical of the fall and winter months and instability is typical of the spring and summer, although this is not to say that the reverse conditions do not occur. Visibilities associated with stable and unstable states of the atmosphere fluctuate accordingly.

## Terrain Effects

The nature of the terrain is more important to visual flight than might at first be supposed. As a rule, the more sparsely inhabited the terrain, the more difficult it is to navigate. If the terrain is flat and featureless as well, this can pose a serious problem. Consequently, visual pilots need to be able to fly higher and pick up more clues as to their whereabouts. Some of these clues may be electronic, in which case reception range is proportional to altitude. The density of navigation aids is also proportional to settlement density. As an example, we can see that the pilot forced down to 1,000 feet above ground by clouds and picking his way through visibilities of 3–5 miles puts himself in a better position by following a familiar highway, shoreline, river, or railroad than that same pilot in the same conditions hundreds of miles from home over featureless prairies or forests.

### MOUNTAINOUS TERRAIN

The need for height and good visibility is even greater in the mountains. One valley looks very much like another, and many valleys wander through all points of the compass. From looking at the chart, the pilot may think that he just has to follow "the valley with the road in it," only to find that all the valleys have roads in them, some built since the map was last revised, others the mapmaker did not see fit to include. Several valleys may look alike but the blind canyons are strewn with the wrecks of aircraft whose pilots took the wrong turn in poor weather.

Weather reporting stations in the mountains are farther apart than in settled flatlands, yet the weather is often local; extreme variations can occur unreported between stations. Weather also can change quickly. Whereas in flat country a string of reports from stations enroute of ceilings of 3,000–4,000 feet and visibilities of 5–10 miles would be decent VFR conditions, the same would not be true for the mountains. Depending on the time of year, area, and time of day, a new pilot unfamiliar with the mountains would have good reason to abandon a flight if ceilings were reported lower than 5,000–7,000 feet, and visibilities less than 10–15 miles.

The mountains produce a syndrome of effects on VFR flight. They produce obscuring phenomena by themselves and amplify other independent weather processes that tend to this same effect. Changes in the weather can occur quickly and locally, yet the reporting stations are far apart. Obscuration to visibility is more frequent and more intense than over flatlands, yet the visual pilot's need for good visibility is also more acute. The new pilot therefore must approach mountain flying with caution and must be prepared to accept a high proportion of cancelled flights, especially in winter.

In the Canadian Rockies whole months can go by in winter with no more than two or three days in each month when VFR flight "across the rocks" is possible. Many FBOs in or near mountain areas will not dispatch flights into the mountains unless the renting pilot has had a "mountain check." On the British Columbia coast a half day of lectures and a whole day of flying in the mountains is considered to be a "mountain check"—which should give some indication to flatland pilots of what is involved.

Hills and mountains provide a continual lifting mechanism for air passing over them. Therefore, more clouds and precipitation will always be present over mountains than over neighboring flatlands. If other lifting mechanisms, such as fronts or convergence are active, orographic lift will be added to them and some wild weather can result. Clouds tend to linger in the mountains after they have dispersed from flatlands because of shade from the sun and shelter from the wind.

## TERRAIN AND NIGHT VFR

The nature of the landscape is sometimes a critical factor in night VFR. Flat, densely settled country provides the easiest flying conditions.

Hilly or mountainous terrain compounds the effect of night illusions. Many airports in these areas are listed in the directories with warnings against night operations by pilots who are unfamiliar with the surroundings. Some pilots will not fly night VFR in the mountains at all. Those who do, other than in conditions of calm, clear, moonlit skies, do so successfully by applying shrewd judgment of local weather conditions, which is matured by ample experience.

Day, night, seasons, and terrain affect the nature and availability of the evidence needed by the new visual pilot to control and navigate the aircraft. These effects are predictable. Pilots must learn to guard against some and turn others to their advantage.

# 12 Obtaining and Using Information

## Sources of varying helpfulness

**INFORMATION** on the present and forecast condition of the sky is abundantly available from a wide range of sources. So much information is available, in fact, that VFR pilots must sort out what is relevant and what is not. It is important to have these sources structured in their minds so that they know which ones to consult and what information to expect.

The most basic source of information is the simple habit of skywatching. As with wildlife or anything else to do with nature, it is astonishing what can be learned by quiet, patient observation. The sometimes mad scramble of modern life is not conducive to this, nevertheless, we should do it whenever we can.

## General Media Forecasts

Radio, television, and newspapers are all good sources of information. Indeed, one early morning television program in the U.S. is specifically about aviation weather. It is sponsored by an insurance company—which should tell us something. The news media tend to give us forecasts for tomorrow's weather in terms of temperature and precipitation. Everyone complains about the inaccuracy of media forecasts, which only goes to show the difficulty of predicting weather 24 hours ahead. We can use these forecasts as a general indication of what to expect, but we should not be blinded to reality by excessive reliance on them, and we need more detailed forecasts for shorter periods nearer in the future. Short-term forecasts valid for the next 6–12 hours often are highly accurate.

The significant feature of a media forecast is "percent chance of rain or snow." It can be cloudy without raining, though it never rains without clouds being present. Moreover, thick, deep, and extensive clouds are required to produce substantial rainfall. If rain or snow is going to fall, the chances are that substantial clouds will be present below 5,000 feet. Nothing is wrong with flying in the rain; it is the conditions producing the rain, and resulting from it, that can be hazardous. We could take the breakeven point as a 50% chance of precipitation. If the chance of precipitation is 50% or greater, the outlook for cross-country VFR is not promising, especially if snow is forecast. If the chance of precipitation is less than 50%, it probably will be a good or reasonable VFR day.

For those of us living in coastal areas, the marine forecasts may be more easily obtainable than the aviation forecasts. The words "small craft warning" in a marine forecast should make a pilot's ears stick up. Wind speeds over the water, however, are generally 10 knots higher than over land, and small-craft warnings generally portend winds rather than low cloud, rain, or fog.

## Aviation Forecasts

Let us now look at the real nuts and bolts of the information we need. This is contained in the aviation forecasts and reports, which will give us detailed, aviation-type news of the weather to be expected on our planned route over the next few hours.

The four main categories of aviation weather report are area forecasts (FA), terminal forecasts (FT), surface actuals or "hour-lies" (SA), and winds aloft (FD). Note that three are forecasts and one is a report of presently observed conditions.

## AREA FORECASTS (FA)

The area forecast covers an expanse of territory that is large in terms of lightplane operations. It probably is several hundred miles across and is divided into regions. We need to know which region of which forecasting area we live in and where the boundaries of the neighboring regions and areas are. Area forecasts are issued every 6 hours about half an hour before their validity periods. Each forecast is valid for 12 hours, with an "outlook" for 12 hours beyond that. Thus, at any given moment, two forecasts are valid, an old one and a new one. Area forecasts are issued at 0000, 0600, 1200, and 1800 Greenwich Mean Time (Z, Zulu), which are different local times in different parts of the world. We should know what times the area forecasts are issued in our own local time.

Suppose that we have an important flight lined up for tomorrow and that we live in an area 5 hours behind Greenwich Mean Time. An area forecast comes out at 6:30 a.m., valid 7 a.m. to 7 p.m., with an outlook up to 7 a.m. tomorrow. The next one comes out at 12:30 p.m., valid 1 p.m. to 1 a.m., and so on. Thus, we cannot even get an outlook for the whole of tomorrow until 7 p.m. this evening, so it is no use trying, although competent weather forecasters give us their best estimates.

The area forecast contains a prognosis ("prog."), which is the big picture of pressure distribution, movement of weather systems, and air-mass types. Here is one for British Columbia:

Upper trough in a north-south line west of Cranbrook at 12Z moving eastwards at about 20 knots. At 12Z an upper ridge in a north-south line along 135° West latitude moving to lie in a north-south line through northern Vancouver Island by 24Z. Surface ridge in a north-south line along 130° West latitude at 12Z moving to lie in a northeast-southwest line from Fort St. John to central Vancouver Island by 24Z. Ahead of the upper trough the air mass is patchily moist and unstable gradually drying and becoming more stable west of the upper trough towards the upper ridge. Moderate northwesterly flow aloft east of the upper ridge becoming westerly in the vicinity of the upper ridge.

That is not easy to interpret, especially when it is read over the telephone. We are likely to get lost trying to scribble it all down, but there is a way around that.

The prognosis is followed by reports for each region in the forecasting area, showing what effects the features and events described in the prognosis will have on that particular region. Let us read on from the prognosis:

> Southern Mountains region. Clouds and weather: east of the upper trough, 6,000 to 7,000 scattered, occasionally broken, tops at 9,000 feet above sea level. 11,000 to 12,000 broken, occasionally overcast, layers to 15,000. High broken with few embedded towering cumulus or altocumulus castellanus, tops at 16,000, giving few light rain or snow showers with visibilities 2 to 5 miles in light snow showers. Local ceilings 400 to 1,500 feet above ground level forming in precipitation with visibilities ¾ to 3 miles. West of the upper trough, 6,000 to 8,000 scattered occasionally broken stratocumulus, tops at 9,000. 12,000 patchily broken, tops at 14,000, high scattered. Stratocumulus becoming cumulus after the middle of the period. Middle or high clouds persisting over extreme southeastern sections giving isolated rain showers except light snow showers at the higher levels. Stratus, fog, and fog and smoke patches forming during the period giving local ceilings 300 to 1,200 feet above ground level and visibilities ½ to 4 miles. Stratus and fog dissipating early in the period.
>
> Icing: light occasionally moderate rime icing in clouds. Moderate mixed rime and clear ice in convective clouds. Freezing level 4,000 to 5,000 feet except for below-freezing surface temperatures in most valleys in the eastern sections until the middle of the period.
>
> Turbulence: moderate in the vicinity of convective clouds.
>
> Outlook: VFR except local IFR because of ceilings and fog mainly in the eastern sections.

That would be fine if we were an airline crew. We would be given a "weather package" with all that in it. In a jet airliner we would cross that region in about half an hour. We would climb or descend through the clouds mentioned, but otherwise we would be cruising above them on an IFR flight plan. As VFR pilots, we typically get a briefing by voice over the telephone. The forecast we have just seen covers the whole gamut of conditions from good VFR down to IFR minima and, for our immediate purposes, it does not really tell us what we need to know. We may want it as backup later in the briefing, but not first thing.

### TERMINAL FORECASTS (FT)

The most important information for us is contained in the terminal forecasts and surface actuals (discussed next). People sometimes have them confused, but they are different in many important respects even though their format is similar. Both are issued at airports and refer only to the weather in that neighborhood. Terminal forecasts are issued only at medium-sized and large airports. Surface actuals are issued for all airports with a Flight Service Station or a control tower, and also at some strategically located spots such as mountain passes. Some remotely located stations are unmanned and automatic.

We tend to assume that the weather changes uniformly between stations. If one place is reporting or forecasting a 4,000-foot ceiling and the next one gives a ceiling of 2,000 feet, we would assume almost unconsciously that the ceiling halfway between would be 3,000 feet. If the stations are 30 miles apart with flat terrain in between, that may be true. If they are 100 miles apart in the mountains, it probably will not be. Even if two stations are 30 or 40 miles apart with a range of hills, a large lake, or a sea strait in between, sky conditions affecting VFR flight quite likely can exist in between without being reported at either.

If two stations 50 miles apart are both reporting clear skies and light winds, the chances of finding weather in between which is 1,000 overcast and visibility 2 miles in light rain and fog, is equivalent to the chances of finding fresh snow in Texas in August. The weather does not work like that. But suppose that, of those two stations, one was reporting or forecasting 2,500 broken, the other 1,000 scattered and 3,000 overcast, and there was a range of hills in between with peaks at 2,000 feet. In that case, we could well have difficulty in getting through the hills.

Knowing which airports issue terminal forecasts is very useful. If Podunk is a 2,000-foot grass strip, no terminal forecast will be forthcoming, and to ask for one will only waste time and diminish the inquirer in the briefer's estimation. It is also important to know at what local times of day the terminal forecasts are issued. They are issued four times a day, not at the same times as the area forecasts: 0500, 1100, 1700, and 2300 GMT. They are valid for either 12 or 24 hours; the validity period is stated at the beginning of each station forecast.

If our local terminal forecasts are issued at midnight and 6 a.m., and at noon and 6 p.m., and if we plan to take off at 8 a.m. tomorrow morning, we can telephone the last thing at night for the 12 p.m. forecast, which will be valid at least until noon tomorrow. It may show us whether getting up that early to go flying is worthwhile! If the forecast is reasonably satisfactory, we have the 6 a.m. terminal forecasts together with the current surface actuals when we get up in the morning. We are then in the strong position of being able to compare two terminal forecasts and to compare the current surface actuals with the most recent terminal forecasts to see how the weather is really developing. If the midnight forecast is moderate, the 6 a.m. update worse, and the surface actuals worse than either, it might be a good reason to cancel the flight there and then. The opposite also would be true.

The ability to obtain this information comes from nothing more than knowing what information is available, and when. We can then ask for the categories we need, and if the briefer knows any other relevant information, he will give it to us. If he knows that he is talking to a well-organized pilot, he is more able and more willing to help than if not. If we come across a briefer who is busy and short-tempered, the very least he can do is to read the terminal forecasts and surface actuals we ask for. That presupposes that we know which ones to ask for and have our pencil poised.

The information in a terminal forecast relates to the area within 5 nautical miles of the center of the airport, with appended remarks, if any, extending to 10 nautical miles. The forecast is divided into time blocks within which conditions are expected to be fairly constant. Within each time block are three pieces of information: clouds, visibility, and wind.

Clouds are always mentioned. If none are forecast, this condition will be described as "clear." Visibility is mentioned only if forecast to be less than 6 miles, in which case the obscuring medium is mentioned as well. Wind is mentioned only if forecast to be stronger than 10 knots, in which case its direction is given in degrees true.

We are interested in ceiling, clouds below 10,000 feet, visibility, and wind speed. If the briefer does not mention the last two, it means that they are respectively more than 6 miles and less than 10 knots. The more common standard abbreviations are worth knowing. Knock the last two 0s off the altitudes, so that 8 stands for 800 and 80 for 8,000. If we write the first digit only, we probably will forget afterward whether it was hundreds or thousands.

Let us try this one. We have asked for the terminal forecast for Kitimat, British Columbia. The briefer reads:

> Kitimat at seventeen Zulu valid for twelve hours. Five hundred scattered, ceiling one thousand broken, two thousand overcast, wind one five zero at ten, occasional visibility four miles in light rain and fog. Becoming at nineteen Zulu two thousand scattered, ceiling five thousand broken, wind one eight zero at fifteen, occasional ceilings two thousand broken, five thousand overcast and light rain showers.

We should already have the place name written down; otherwise we will still be doing so when the briefer is halfway through the forecast. This is what we write as he reads:

17Z  5⦵  C10⦵  20⊗  150/10  OCL  4R-F
19Z  20⦶  C50⦵  180/15  OCL  C20⦵  50⊗  RW—

From the operation of Murphy's Laws, it seems that any bad sky condition forecast as "variable," "occasional," or "risk" is almost certain to occur and we had better make allowances accordingly. Kitimat happens to be in the mountains in a part of the Pacific coast not noted for its fine weather, and this particular forecast, although improving, is not a good VFR situation for that area. Floatplane pilots operating along the coast would regard it as satisfactory, but they fly routinely in far worse conditions than those acceptable to a new landplane pilot.

## SURFACE ACTUALS (SA)

The surface actuals are laid out differently from the terminal forecasts. They are issued every hour on the hour. "Specials" may be issued at any time when the weather changes significantly. A "Regular Special" is a Special that happens to be issued on the hour. A surface actual is a complete report of the weather as observed at one time from one place on the ground. We hear them all the time in ATIS broadcasts. After the name of the station and time of the report, the following information comes out in this order:

1. Clouds
2. Visibility—always reported whatever it is
3. Obscurations—reported only if present
4. Barometric pressure in millibars
5. Temperature and dewpoint—°F in the U.S., °C in Canada
6. Wind—always reported, degrees true and knots
7. Altimeter setting
8. Cloud types—types and tenths sky coverage layer by layer up to the highest layer visible
9. Remarks and pressure tendency

The network of stations issuing surface actuals is denser than for terminal forecasts. Most of them are at airports, but others are situated elsewhere in more remote areas, and these are well worth knowing about.

Here is a surface actual as it would be read:

Campbell River at eighteen Zulu. Five hundred scattered, balloon measured ceiling eight hundred broken, two thousand overcast, visibility four miles in light rain and fog, pressure ten oh four point four millibars, temperature six, dewpoint six, wind one zero zero at five, altimeter two niner six five, stratus fractus two-tenths, stratocumulus six-tenths, stratocumulus two-tenths, visibility to the north-west two miles in fog.

We write:

$$18Z \quad 5\oplus \quad BB\oplus \quad 20\ \varnothing \quad 4R-F \quad 1004.4 \quad 6/6 \quad 100/5 \quad 2916$$
$$SF2 \quad SC6 \quad SC2 \quad VIS\ NW \quad 2F$$

We should be quick to notice the danger signal contained in the fact that the temperature and dewpoint are the same. The situation is conducive to formation of low clouds and fog, which are reported already. Campbell River is on a coastal plain with hills to the west and an open strait to the east. Floatplanes and helicopters probably will be operating visually in these conditions, but it is a thoroughly bad situation for a new visual pilot.

## WINDS ALOFT (FD)

Winds aloft are forecast at points still more widely scattered than terminal forecast stations. It is useful to know where the forecasting stations are; if we ask for winds aloft for some place that does not issue an FD, we will waste time. The relevant times here are the "valid times," which are 0000, 0600, 1200, and 1800 GMT. The valid periods are 3 hours before and after each valid time. If the station is more than 1,500 feet above sea level, it will give no forecast wind for the 3,000-foot level. No temperatures are forecast for the 3,000-foot level or for any level within 1,500 feet of station elevation. The winds aloft forecasts for Denver, for example, would begin at 9,000 feet, and a temperature would be given. Seattle would give a 3,000-foot wind but no temperature. We should not waste time by asking for information that is not given.

## ADDITIONAL INFORMATION

*Pilot Reports* (PIREP/UA). These reports are valuable because they give the only true picture of what conditions really are.

If we meet with conditions that force us to turn around, or that we would not like someone else to experience, we should transmit a PIREP to Flight Watch on 122.0 MHz in the United States or on 126.7 MHz in Canada (or by telephone after landing, if appropriate). What they need to know is that we are a VFR aircraft, what conditions we encountered, and when and where we encountered them. If the recipient wants any other information, he will ask.

*SIGMET.* A SIGMET is issued to warn pilots of violent weather. Fog, for example, does not merit a SIGMET. They are issued for severe thunderstorms, tornadoes, violently active fronts, severe icing or turbulence, hurricanes, and the like. If a SIGMET is in force for our area, the briefer will almost certainly tell us. Whether we wish to enter or not enter an area covered by a SIGMET depends on the nature of the weather that caused the SIGMET to be issued, the terrain, the time of day, and how good we judge our understanding of the weather situation to be. If a SIGMET was issued for a cold front, we had seen the cold front pass, and now the strong winds are abating under a clear sky, conditions might be safe for us to fly. On the other hand, if it was midafternoon and a SIGMET was issued for thunderstorms in a mountainous area, we would have good reason for avoiding flying in the area at all.

*Transcribed Weather Broadcasts* (TWEB). TWEBs are transmitted continuously from certain NDBs and VORs. The transmitting stations are identified on the aviation charts.

*Weather Surveillance Radar.* This service operates over much of the U.S. and parts of Canada. It detects precipitation, which is used to indicate thunderstorm activity.

Aviation weather offices display weather charts, sometimes in bewildering profusion. The "surface" charts are the ones most relevant to the visual pilot. A useful chart sometimes available is the plot of recent lightning strikes. This provides a valuable indication of where thunderstorms are and how they are arranged, whether isolated, densely packed, or in lines.

*Notices to Airmen* (NOTAM). NOTAMs are of use mostly to IFR pilots because they warn of navigation aids being out of service. For VFR pilots they warn of things such as runways being closed, aerial firefighting, air shows in progress, and other important safety information. They are mentioned here because they can be obtained with a weather briefing — and should be.

## BOOKS

One source of general weather information that should be mentioned is books. For a start, a new pilot should read a good selection of aviation weather books. It is also well worthwhile to browse in the general meteorology section of the local public library. There are some exceptionally well-produced books on both general and marine meteorology, and they tend to be better illustrated than the books produced for the relatively smaller aviation market. Many of these books give insights that cannot be gained from the generally repetitive aviation weather books. It is no overstatement to say that the new pilot's life can depend on a correct interpretation of the weather, and no avenue for self-education should be left unexplored.

## Putting the Information to Use

All this information is fine, but how do we put it together? How do we extract the right information from the system at the right time and then process it to come to the right decision? We might as well have a procedure for weather-planning a flight. The following one seems to work.

On the day before the flight, we can gain some idea of the probable weather. Perhaps we see a weather map in a newspaper or on television; perhaps we hear a weather forecast on the radio. Before we go to bed a set of forecasts may have been issued already, with a validity period extending over the intended duration of the flight. Last, we should look outside. If the wind is blowing more strongly than a breeze, there is a fair chance that different weather will greet us in the morning. But, when all is said and done, it is mostly speculation and we cannot do anything about it anyway.

The first thing we do on getting up is to look out the windows — on both sides of the building, if possible. We then telephone the weather office. It does not even have to be the nearest one. If there is one a hundred miles away with an 800 number and the operator gives good briefings, we should use it. The best place is wherever we can get a human being on the end of the line who knows how to give a weather briefing. But first we must know what we are going to tell the briefer and what information we want in return. Many briefers are not pilots and do not know what our real concerns are.

What the briefer needs to know is:

1. Type of operation—VFR, single-engine IFR, oxygen-equipped, multiengine IFR
2. Date and time of flight
3. Origin and destination
4. One-way or return

The sequence we really need is:

1. Terminal forecasts for the route
2. Current surface actuals for the route
3. Winds aloft for 3,000, 6,000, 9,000 feet
4. What is causing it all? (Area Forecast)
5. Relevant NOTAM

If Items 1 and 2 show that the whole route is clear with light winds, or that it is unflyable, we can short-circuit the whole procedure. We can save the briefer's time by zeroing in on a few key issues or by thanking him and hanging up. The more complicated the situation, the more detail we need.

We are several steps ahead of the game if we know in advance which stations enroute issue terminal forecasts, surface actuals, and winds aloft. We should have a piece of paper and a pencil ready, with the stations and report types already written down.

Sometimes it is obvious whether we can or cannot fly the route. Often, however, it is like the sky—a shade of gray. In that case, we need to analyze things in peace and quiet before leaving for the airport, and to decide on the important features for later investigation. Let us listen in on a briefing:

(It is 7:30 on an April morning. The weather, as seen outside the window, looks good.)

*"Good morning. Vancouver Flight Service."*

"Good morning. I'm looking at a VFR flight today, Victoria to Port Hardy about midmorning, returning this afternoon. Could you give me the terminal forecasts for Victoria, Comox, and Port Hardy, please?"

*"Victoria and Comox look pretty good all day, but Port Hardy is calling for nine hundred scattered, ceilings of four thousand broken, occasional ceilings nine hundred broken, one and a half miles in light rain and fog. Weather right now is measured twenty-six hundred broken, visibility twenty miles. Alert Bay just down the coast a ways is reporting a nine-thousand-foot ceiling with some lower clouds between one thousand and two thousand."*

(This requires closer investigation.)

"What's the big picture?"

*"Generally we have a front from Port Hardy to Cape Saint James gradually moving in on the north tip of Vancouver Island. Other than that, there's not much happening."*

"Any applicable NOTAM?"

*"Let's see now. No, nothing."*

"OK, thank you very much."

*"You're quite welcome, sir. Good day."*

Port Hardy is at the north end of Vancouver Island, just off the coast of British Columbia. It is not noted for good weather; April is an uncertain month in that part of the world. For that reason, we are particularly interested in quiet conditions which are not going to change much.

The southern part of Vancouver Island—home territory—is safe all day. The problem is the worsening weather to be expected at Port Hardy as the front moves in. We might be forced to turn around and retreat; or we might make it there, and then be weathered in. On the other hand, we have been there before. The airport is on the edge of a sea strait with an unobstructed approach along the shoreline. We would be flying into the leading edge of a front, which is fine provided that we know when to turn around. There will be fine settled weather behind us all day, as the front is not moving very fast. It looks as if we might decide to go.

On reaching Victoria Airport at 10:30, we obtain a further briefing and decide whether to take off or not.

*"Good morning. Aviation weather."*

"Good morning. I'm looking at a VFR flight from Victoria to Port Hardy this morning, returning this afternoon. Could you give me the terminal forecasts for Victoria, Comox, and Port Hardy, and the current surface actuals for Campbell River and Port Hardy?"

Campbell River is the last airport before Port Hardy and is separated from it by nearly a hundred miles of straits, channels, mountainous islands, and forested mountain country. It is important to make sure that all is well there; it does not issue a terminal forecast. Alert Bay is so close to Port Hardy that weather conditions there will be similar.

*"Victoria is forecasting twelve thousand scattered, high scattered up to oh five Zulu. Comox up to seventeen Zulu tomorrow, six thousand scattered, ten thousand scattered, high scattered. Port Hardy to oh five Zulu, three thousand scattered, ceiling ten thousand broken, occasionally seven hundred scattered, ceiling twenty-five hundred overcast, one and a half miles in light drizzle and fog. Present weather at Port Hardy is five hundred scattered, measured ceiling twenty-six hundred broken, ten thousand overcast, twenty miles, temperature nine, dewpoint seven, wind one hundred degrees at four knots, altimeter three zero two zero, stratus fractus two-tenths, stratocumulus seven-tenths, altocumulus one-tenth. Campbell River, six thousand scattered, thirty thousand scattered, fifteen plus, cumulus one-tenth, cirrus one-tenth."*

"Winds at three, six, and nine, please?"

*"Vancouver at three thousand, light; at six thousand, two-ninety at six and plus three; at nine thousand, two-eighty at fourteen and minus one. Port Hardy at three thousand, two-twenty at thirteen; at six, two-thirty at nineteen, zero; at nine, two-forty at twenty-seven and minus three."*

"What's happening to the front?"

*"It's weak and stationary."*

"Any applicable NOTAM?"

*"Just one sec. No, none."*

"Thank you very much. Good-bye."

The weather close to home is perfect. The front is there but is not in any way threatening. Although some lower clouds have appeared at Port Hardy, the most recent terminal forecast is more favorable than its predecessor. There is no significant change in the surface actuals in 3 hours (between the 7 a.m. one gathered at 7:30 and the 10 a.m. one gathered at 10:30). Both the surface actuals are better than the first (and worst) terminal forecast, and in line with the newer one. The weather is settled, so we take off.

An hour into the flight, near Campbell River, we obtain a further minibriefing. The one thing we want to know now is the Port Hardy weather.

"Campbell River Radio, Cessna one-eighty-two Golf X-Ray Mike Yankee with you on one-twenty-six-seven."

*"X-Ray Mike Yankee, Campbell River, go ahead."*

"X-Ray Mike Yankee, do you have an update on the Port Hardy weather, please?"

*"X-Ray Mike Yankee, stand by."*

*"X-Ray Mike Yankee, Campbell River, the Port Hardy weather on the last hour was reported at fifteen hundred scattered, estimated ceiling thirty-eight hundred overcast, wind two hundred at three, temperature niner, dewpoint six, altimeter three zero one seven."*

"X-Ray Mike Yankee, roger. Thanks a lot."

The front appears 50 miles from Port Hardy as a shelf of stratocumulus with its leading edge at 4,000 feet. We fly down the strait beneath it as it lowers to about 3,000 feet over our destination. There is no wind, no rain, and the journey is uneventful, as is the return trip to Victoria.

Obviously, the weather can assume an infinite variety of forms, and nothing is to be gained by describing further specific situations. It is hoped, however, that this chapter will have provided some ideas on a usable weather-planning procedure.

# 13

# Aircraft and Pilot

## But how good is the pilot?

**TO THE VFR PILOT** everything to do with weather might seem to favor the highly trained instrument pilot with a high-performance aircraft. In terms of arriving at a destination with any degree of reliability, this is true. On the other hand, the light aircraft does have some marked advantages for certain types of operation, and VFR pilots do have certain cards in their deck, which they must play to their fullest advantage.

## Fuel Reserves

Most VFR flights are of less than 3 hours' duration. Most light aircraft have fuel tanks that will allow them to fly for 5 or 6 hours. A Cessna 152 with long-range tanks actually can remain airborne for nearly 9 hours if flown at no more than 45% power—about 2,000 rpm. Therefore, compared to a commercial aircraft, lightplanes have the capacity for massive fuel reserves. Airliners seldom, if ever, take off with twice as much fuel as they

**155**

need to accomplish their intended flight, yet lightplanes often do so.

Many light aircraft offer the pilot the options of full loads of either passengers or fuel but not both. When the aircraft is flown without a full passenger or baggage load, the tanks should be kept as full as possible, unless there is some specific reason for taking off light. Typically, if pilots take off with full tanks, they have the capability to fly almost the whole way to their destination, turn around in the face of foul weather, and fly all the way back to the starting point without having to worry about fuel. They are able to loiter for an hour or more waiting for fog to clear, surface winds to drop, or thunderstorms to die out or move away.

Because the aircraft is moving through the air at high speed, pilots have a natural tendency to feel that everything has to be done in a hurry. Many people ruin perfectly good situations by doing something when, in fact, they should be doing nothing. Flying around burning up fuel is pointless if a preferable alternative exists; but it is essential that the tanks contain enough fuel to carry out that alternative.

Nevertheless, could not the pilot who damaged the aircraft landing in a 30-knot wind have flown around the airport for a while on the chance that the wind might drop to a speed he was better able to handle? Could he not have found another, more sheltered, airport? If he did not carry enough fuel, of course, the answer would be "No." Light aircraft have far more fuel capacity in relation to their average duration of flight than do larger aircraft. It is a real advantage, which should be used.

## Flexibility

The light aircraft is flexible in performance. If handled properly, it can fly very slowly. It can fly about in confined spaces. It can operate from airstrips that the pilot of a larger aircraft could not even consider. It can even operate from unprepared surfaces.

One pilot was flying a Cessna 152 up a mountain pass only to find the pass filled with clouds. On turning around, he found that his retreat was cut off by snow showers in the valley through which he had just flown. (It is the thesis of this book that, if he had been truly weather-wise, he would not have been there in the first place, but, never mind—there he was.) He spotted a half-built section of highway below him. After making no less than half a dozen circuits for inspection and trial approaches, he landed on it, waited for the weather to clear, and went his way.

Another pilot was caught in a similar situation in the same pass. He was not so skillful. He made no trial approaches and landed in a tailwind. He ran into a concrete barrier and damaged the aircraft beyond repair.

It is easy to climb into the blue sky, set cruising power, and fly fast and straight. What is less easy is to fly slowly, yet precisely, to creep along a cloudy valley at 60 knots, and, when things turn really bad, to fly accurate minimum-radius turns in turbulence over irregular, sloping ground and land safely on an unprepared surface. The higher the aircraft's cruising speed, the more necessary it is to practice flying slowly.

## Aircraft Performance

One of the VFR pilot's best defenses against the vagaries of the weather is the performance of his aircraft. It follows that he must know and be able to use every scrap of that performance.

Most of us do not have many options as to the performance of the aircraft we fly. If the local FBO rents a certain type of aircraft, that is what we are stuck with for better or worse, unless we decide to go into partnership or ownership, which tends to be expensive. If we did have some say in the matter, what would we choose for a "weather-proof" aircraft? How about *speed, power,* and *mass?* Unfortunately, they all cost money.

### SPEED

The only certain thing about the weather is that it will change. The main questions are: how fast, and how soon? Besides the weather, we also have no control over the duration of daylight. With the onset of night, our requirements for ceiling and visibility increase, and at the same time those same features often diminish. VFR conditions are sometimes available for only short periods of time. How far can we fly in that time? Certainly we know that feeling. We have planned an all-day trip. The day dawns rainy and foggy but likely to clear. Before long we are chewing our fingernails, wondering if enough daylight remains for us to fly the trip. With a fast aircraft we can afford to "sit the weather out" and then move rapidly when it is good.

An aircraft cruising at 130 to 150 knots has capabilities, in terms of getting the job done, that a slower aircraft, cruising at perhaps 100 knots, does not. Speed means that we can choose from a selection of routes to find the most favorable weather. A few extra miles does not take significantly longer, yet the indirect route may provide better weather.

Speed minimizes the effect of headwinds and maximizes the effect of tailwinds. The pilot of a low-powered aircraft butting into a stiff headwind has much more to worry about in terms of fuel reserves, the movement of weather systems, and the onset of darkness, than does the pilot of a faster one. Suppose, too, that darkness will be upon us in half an hour and we want to be on the ground by then. In still air we can travel 50 miles in that time at 105 knots, 65 miles at 135 knots. That may not sound like much of a difference but, if we can fly in any direction from our present position, that extra speed will give us a 66% greater territory in which to find a place to land.

## POWER

Power confers the ability to handle high density altitudes. In the high terrain of the western parts of North America, the power to take off at high density altitudes and to climb to great heights is a real asset. Flight at higher altitudes not only makes navigation in the mountains easier, but it also allows smoother rides along more direct routes. Foul weather can be seen or heard about on the radio from that much further ahead.

## MASS

Mass is another asset because a heavier aircraft gives a better ride in turbulence and handles better in high winds near the ground. To overturn a 3,000-pound aircraft on the ground takes a very strong wind indeed. The lightplanes that figure in the accident statistics by being flipped over by wind (or occasionally by jet blast) are always in the lighter end of the size range.

## The 180° Turn

Not least among the skills necessary to the new pilot's survival is the knowledge of when to turn around. The decision often is made under considerable stress with limited information available. The desire to reach the intended destination — to "get through," to "tough it out" — generates its own pressure to continue, regardless of how important or how trivial the reason for the flight may be. The combined situation of weather, light, and terrain may turn against the pilot only gradually. The self-deception that conditions may improve, or not get any worse, comes easily to the mind.

The weather briefing must have been at least reasonably in the pilot's favor; otherwise the flight would not have launched. Somehow, making the transition from the information in the weather briefing to the reality surrounding the aircraft is difficult.

VFR is a realm of flight whose specified limits of visibility are well below what any new pilot should consider safe for a cross-country flight. Consequently, it is difficult to suggest any definite conditions which should prompt the new pilot to turn around and retreat.

It can be said, however, that if visual pilots are forced by clouds to an altitude lower than 1,000 to 1,500 feet above flat terrain, and if visibility deteriorates below 3–4 miles, they are entering an area of considerable risk and uncertainty and would be well advised to retreat or find an airfield and land. These figures apply to day VFR over flat terrain. At night they probably should be tripled and, in the mountains by day, quadrupled. The 180° turn has to be made in good time because weather may be deteriorating behind as well as ahead. The following are definite danger signals:

1. Precipitation, especially snow, beginning
2. Turbulence increasing under cloudy skies
3. Sky darkening because of clouds
4. Cumulonimbus anvils, lightning, increased radio static
5. Rapidly falling altimeter settings
6. Terrain and cloud ceiling converging
7. Low scud or ground fog coming into view
8. Any of the above toward nightfall

With regard to the weather, the visual pilot must learn to accept that "what you see is what you've got." The pilot is trained to obey rules. The rules and the information pertaining to them come from the omnipotent, omniscient "Them." Suppose we were to receive a weather briefing from Them, indicating acceptable VFR conditions in the area of our intended flight. Well into the flight we find conditions worsening around us. From the ingrained habit of obedience and acceptance of Their dicta as The Ultimate Word, we cannot at first believe what we see. We try to reconcile this contradiction by thinking that the conditions we see can only be local or temporary. However, no weather report, except a PIREP, is anything more than what someone on the ground said he saw at some particular time. The only real weather is what we see around us. We are the sole and final judges of whether we can fly safely in what we see ahead; "What you see is what you've got."

The visual pilot who does not turn around soon enough may end up in clouds at low altitude and uncertain of his whereabouts. The danger of this situation cannot be overstated, and the outcome is often fatal. All IFR operations less than 2,000 feet above ground level are confined to the departure and approach phases of the flight, during which the aircraft must be flown with precision along published tracks and not below certain stated altitudes. It should be obvious that the visual pilot blundering about at low altitude among clouds, rain, hills, and obstructions is in a position of extreme hazard from which he may not extricate himself in one piece.

Ability to control the aircraft by instrument references alone is a skill that lapses with disuse. A few hours of dual instruction under the hood in calm, clear skies is poor preparation for unintentional entry into the turbulent interior of a cloud.

Safe visual flying results from the successful interaction of the pilot and the aircraft with the weather. The weather is beyond our control and sometimes beyond our ability to predict. Almost all light aircraft flying today are of sound construction and in airworthy condition. We must ensure that the pilot does not become the weak link in the chain.

# 14 What About an Instrument Rating?

## Getting your wings wet

**A BOOK OF THIS NATURE** would not be complete without an insight into the benefits of obtaining an instrument rating and the steps involved in doing so.

The new private pilot's idea of IFR operations, which is gained from books and magazines, runs something like the following. The instrument pilot takes off into a low overcast, and flies continuously on instruments for hundreds of miles, weaving between thunderstorms, trying to avoid or beat off ice, copying and following clearances of incredible complexity. The pilot then shoots an equally complicated approach procedure and sneaks in to a landing out of a 200-foot overcast in heavy rain and a high wind.

Most people are scared off by what they think are typical IFR operations. It is true that, if instrument pilots so desire, they can go and find the conditions just described. If they are working for an air carrier, it will be their business to fly in those conditions if need be and, moreover, to do so at night. For instrument-rated private pilots, this is not true. Just as in VFR flying, they pick their weather. In doing so, they will find that a wide variety of weather conditions that led to cancelled flights or white-knuckle specials under VFR become safe and easy under IFR.

One November Saturday morning on the coast of British Columbia, three friends decided to make a 45-minute flight to a nearby airport, have lunch, and return to home base. One pilot was instrument-rated; the other two were not. Conditions were VFR in the morning but likely to deteriorate to marginal VFR in the afternoon. The flight was over water studded with mountainous islands, and the three pilots decided they would not make the flight without instrument capability. They rented an IFR-equipped Cessna 172 and agreed that one of the visual pilots would fly the outbound leg. The other would fly the homeward-bound leg unless conditions had deteriorated to the point at which IFR was believed to be the safer option, in which case the instrument-rated pilot would fly home.

On the outbound leg they were, in fact, unable to reach their objective because of a belt of very low clouds along a shoreline. Visibility while approaching the shoreline was 10–15 miles under an overcast at 4,000–5,000 feet. The belt of low clouds, which they could see dimly through rain falling from the overcast, seemed to be about 1,000 feet thick, based about 500 feet above ground level; only small patches of land were visible beneath it. As they entered the rain, visibility shrank to about 3 miles and horizon reference could be maintained only by looking back into clearer air, where a peninsula could be seen. Seeing that they were entering uncertain conditions in which loss of horizon seemed imminent, the pilot diverted to another airport in better visibility, where they landed and had lunch.

An hour and a half later, conditions had deteriorated to a ceiling of 1,000 feet overcast, visibility 2 miles in rain and fog, with a 15-knot wind. Conditions at home base were reported to be good VFR. They decided that attempting to return to base under VFR could be dangerous and certainly would not be enjoyable. The instrument pilot therefore filed an IFR flight plan.

The departure runway ended on a shoreline, and outside horizon reference was lost immediately on takeoff over the water. They climbed to 4,000 feet and cruised in clouds for 20 minutes before breaking out into good VFR conditions, in which they made a visual approach to home base. The whole flight was simple, safe, and convenient. It was completed with a total of four short clearances.

This could be said to be a trivial example. What was not trivial was the fact that at about the time when the three friends were landing at home base, a pilot without instrument training, although flying a fully instrument-equipped aircraft, became disoriented in the fog and crashed fatally within a few miles of the airfield the three friends had just left.

## Requirements

The path from private license to instrument rating is not simple; it is not easy; it is not cheap. It requires an investment of time and money equivalent to that needed for the private license all over again. Much of the basic instrument training can be done in any aircraft fitted with the basic blind-flying panel. Some of the training, both basic and advanced, can be done in a simulator. Some of the advanced training will have to be flown in an IFR-equipped aircraft, which may rent for a higher rate than one with less equipment. These offsetting factors should result in an average rental cost no higher than that for the private license.

The candidate cannot be issued an instrument rating until he has accumulated pilot-in-command time of 150–200 hours, depending on government regulations. A portion of this time must be spent cross-country flying, perforce under VFR.

## Stages of Training

Five basic steps underlie the training for the instrument rating. First, the pupil has to learn to control the aircraft by instrument references alone. Training takes place in clear air, but the pupil wears a "hood" like a baseball cap with an enlarged peak, which shuts off the view of the outside world. The instructor has an unobstructed view, watches for traffic, and assigns headings and altitudes to fly. The pupil experiences and learns to suppress the sensory illusions referred to in Chapter 5.

This training forms part of the private-pilot curriculum in the United States and may do so in Canada in the future. Instrument flying "under the hood" is only a simulation of the real thing; the simulation can be much improved by flying at night.

Next, the pupils must learn to fly more accurate headings and altitudes by instruments than they needed to under VFR. This stage includes not only level turns and straight climbs and descents, but also climbing and descending turns in which the required heading is reached before, at the same time as, or after the required altitude. Maneuvers begin and end on precise headings and altitudes. The pupils must also learn to fly "partial panel" as if the attitude and heading indicators were inoperative. In this and other stages, work in the aircraft may parallel work in the simulator, according to the facilities available and the instructor's inclinations.

Once they can fly the aircraft by instruments with reasonable precision, the pupils learn to fly precise tracks with reference to ground-based navaids, VOR, ILS, and NDB. Use of these navaids is preliminary to learning departure, enroute, and approach procedures. Approaches receive more attention than do departures. An instrument approach consists of flying a series of exact tracks and altitudes so that the aircraft will avoid obstructions during its blind descent and will break out of cloud in a position to make a visual landing.

This phase of training also is carried out in the simulator and under the hood. Some traffic control facilities allow the instructor to file an IFR flight plan for "consecutive approaches," which enables some training to be carried out in IFR conditions. The pupils can also practice for themselves in VFR conditions under the hood, provided that a safety pilot (who does not have to be an instructor or instrument-rated) is riding with them, and that an understanding is reached with ATC that the flight is "simulated IFR." The ATC clearance, if given, will contain the instruction to "maintain VFR at all times." In case of dense traffic, the necessary clearances may be refused or the pupil may spend more time being vectored around traffic than learning approaches.

The final stage of training consists of flying complete IFR flight plans. These exercises include missed approaches. Such exercises can be "flown" both in the simulator and in the aircraft. It is most important that the trainee experience actual IFR conditions, including strong winds, ice, and turbulence. Candidates not uncommonly complete their training without ever having flown in clouds. Some people, accustomed to flying under the hood on sunny days, become frightened and confused when they find themselves inside a turbulent cloud, perhaps picking up a trace of ice as well, for the first time without an instructor.

The instrument rating is awarded after passing a rigorous written examination and flight test. Flying schools run IFR groundschools at intervals depending on the size of the school. The instrument trainee is not allowed to file an IFR flight plan, even in perfect weather, until after passing the IFR flight test. Therefore, the flight test is the candidate's first IFR flight plan without the direct supervision of an instructor. This can have curious and unnerving effects.

## The New Rating

New instrument pilots will not perform as well on their first IFR flight as they did while under instruction. It is as if a private pilot were not allowed to fly solo until after passing the private-pilot flight test. Therefore, the newly rated pilot would be wise to start out by flying IFR in conditions of ceiling and visibility low enough to necessitate flying solely by instruments for an appreciable part of the journey, but which do not demand tough decisions or a missed approach at destination.

Instrument-rated pilots will find that IFR capability brings peace of mind to all sorts of marginal-VFR situations, as well as enabling them to fly in actual IFR weather. They will experience a sudden and astonishing decrease in cancelled or diverted flights, although, on occasion high winds, ice, thunderstorms, and very low ceilings or fog will keep them on the ground. It will be their own decision what conditions they wish to handle at night. Some airports they may be in the habit of using may not have instrument approaches. Single-engine IFR over high mountains will remain a questionable activity.

In general, however, with IFR capability, weather problems will evaporate like morning mist. At the same time, the pilots' ability to interpret weather will advance by leaps and bounds. One reason for this is that they will be directly experiencing weather that would have grounded them under VFR.

New VFR pilots are hampered and sometimes endangered by conditions of light, ceiling, and visibility which occur frequently and for extended periods of time. New pilots' ability to judge their effects and predict their likely extent and duration is in its infancy. If, after building up the requisite pilot-in-command time, the pilot qualifies for an instrument rating, the atmospheric conditions adversely affecting the flight will be found to be fewer, less frequent, and of shorter duration. At the same time, the ability to interpret them is better developed. The result is a major advance in the safety and reliability of operations and in their credibility as pilots.

# CONCLUSION

**WHEN WE LAUNCH OURSELVES** into thin air, we enfold ourselves in and rest our weight upon a fluid whose processes are in the highest degree mysterious because they can be measured, for the most part, only indirectly and observed only by their results. The restless ocean of the sky, though often placid, is not a forgiving environment. The new pilot's ability to interpret and predict its motions and changes is meager; the pilot and his tiny, feeble aircraft are playthings in the grip of its immense power and its endless capacity to blind, confuse, and delude.

Weather makes liars of us all, which is why this book is written in the words of uncertainty — may, might, could, often, sometimes. The violence and the gentleness of the sky take us equally by surprise. We can predict the motion of the stars centuries in advance, yet we are hard put to forecast the behavior of our own atmosphere beyond tomorrow.

In spite of this, or perhaps because of it, new pilots must miss no opportunity to increase their understanding of the sky. They must learn to assess shrewdly, both before takeoff and while airborne, whether they are likely to enjoy the sky's continued favor or whether it is turning against them. If the latter, they must either stay on the ground or retreat and find a safe haven. No extraneous pressures must be allowed to influence these decisions, because to do so is dangerous. Whatever the transportation needs of the pilot and the passengers, they must be shelved or met in some other way.

The only certainty about the sky is that it will change. It is tragic indeed when the sun shines down from a clear sky on the wreckage of an aircraft that crashed in storm and darkness only a few hours before.

New pilots will not win a showdown with the sky. The purpose of this book is to explain those features of the sky that can threaten the safe and pleasant conduct of a flight and why they tend to do this. Knowledge of these features will enable pilots to guard themselves, their passengers, and their aircraft for a lifelong enjoyment of the sky.

# BIBLIOGRAPHY

Atkinson, B. W., and Gadd, A. *Weather.* New York: Weidenfeld & Nicholson, 1987. (ISBN 1–55584–028–0)

Benstead, C. R. *The Weather Eye.* London, U.K.: Robert Hale, 1954.

Buck, R. N. *Weather Flying.* New York: Macmillan, 1978. (ISBN 0–02–518020–7)

Byers, H. R. *General Meteorology,* 3d ed. New York: McGraw-Hill, 1959. (LOC# 58–59658)

Cagle, M. W., and Halpine, C. G. *A Pilot's Meteorology,* 3d ed. New York: Van Nostrand Reinhold, 1970. (ISBN 0–442–21435–9)

Collins, R. L. *Flying IFR.* New York: Macmillan, 1978. (ISBN 0–02–527190–3)

_____. *Flying the Weather Map.* New York: Delacorte Press, 1979. (ISBN 0–440–02610–5)

_____. *Thunderstorms and Airplanes.* New York: Macmillan, 1982. (ISBN 0–02–527250)

Crawford, W. P. *Mariner's Weather.* New York: W. W. Norton, 1978. (ISBN 0–393–03221–3)

Day, J. A. *The Science of Weather.* Reading, MA: Addison-Wesley, 1966. (LOC# 66–22573)

Dunlop, S., and Wilson, F. *The Larousse Guide to Weather Forecasting.* New York: Larousse, 1982. (ISBN 0–88332–288–9)

Environment Canada. *Weather Ways,* 3d ed. Ottawa: Canadian Government Publishing Centre, 1982. (ISBN 0–660–11109–8)

Geiger, R. *The Climate Near the Ground.* (M. N. Stewart et al., trans.) Cambridge, MA: Harvard University Press, 1957 and subsequent editions. (Original work published in German, 1927)

Guerny, G., and Skiera, J. P. *Pilot's Handbook of Weather,* 2d ed. Fallbrook, CA: Aero Publishers, 1974. (ISBN 0–8168–7355–0)

Hardy, R., Wright, P., Gribbin, J., and King, J. *The Weather Book.* Toronto: John Wiley, 1982. (ISBN 0–471–79877–0)

Imeson, S. *Mountain Flying.* Long Beach, CA: Airguide Publications, 1982. (LOC# 82–716983)

Kotsch, W. J. *Weather for the Mariner.* Annapolis, MD: Naval Institute Press, 1983. (ISBN 0–87021–756–9)

McCollam, J. *The Yachtsman's Weather Manual.* New York: Dodd, Mead, 1973. (ISBN 0–396–06721–2)

Neiburger, M., Edinger, J. G., and Bonner, W. D. *Understanding Our*

*Atmospheric Environment.* San Francisco: W. H. Freeman, 1982. (ISBN 0–7167–1348–9)

Schaefer, V. J., and Day, J. A. *A Field Guide to the Atmosphere.* Boston: Houghton Mifflin, 1981. (ISBN 0–395–24080–8)

Schiff, B. *The Proficient Pilot.* New York: Macmillan, 1985. (ISBN 0–02–607150–9)

———. *The Proficient Pilot II.* New York: Macmillan, 1987. (ISBN 0–02–607151–7)

Taylor, R. L. *Instrument Flying.* New York: Macmillan, 1972. (ISBN 0–02–616670–4)

Wallington, C. E. *Meteorology for Glider Pilots,* 3d ed. London, U.K.: John Murray, 1977. (ISBN 0–7195–3303–1)

Welch, A. C. E. *Pilot's Weather: A Flying Manual.* New York: Frederick Fell, 1973. (ISBN 0–81190–288–9)

Welch, J. F. (Ed.). *Van Sickle's Modern Airmanship,* 5th ed. New York: Van Nostrand Reinhold, 1981. (ISBN 0–442–25793–7)

U.S. Air Force. *Weather For Aircrews.* Air Force Manual 51-12, 1974.

# INDEX